5th Spring 2009.

Party Wall etc. Act 1996

Misund[...]ndings

G[...]

A M Frame MSc., MRICS., FFPWS., FASI., FCIOB., [...]
Chartered Building Surveyor

Published by
FACULTY OF PARTY WALL SURVEYORS
P.O.Box 86
Rye
TN31 9BN
01424 883300 – enq@fpws.org.uk
www.fpws.org.uk

First Edition 2007

First Published in England
by
Faculty of Party Wall Surveyors

Acknowledgements:
The Directors of the Faculty of Party Wall Surveyors.
Stuart Frame – Barrister at Law.
Party Wall etc. Act 1996 is reproduced under the terms of Crown Copyright Policy Guidance issued by HMSO.

Cover Photograph – Hadrian's Wall by Graeme Peacock of Photographicimagery, 5 Moore Court, Newburn, Newcastle Upon Tyne. NE15 8QE

Party Wall etc. Act 1996

MISUNDERSTANDINGS

and

GUIDANCE

Alex M Frame

Contents

INTRODUCTION

The Party Wall etc. Act 1996 has now been in operation since 1st July1997, although in its previous form as Part VI of the London Building Acts (Amendment) Act, 1939, where surveyors had previously used it for very many years. The jurisdiction of the London Building Act was of course only for the Inner London Boroughs, but now this new Act is applicable to the whole of England and Wales.

The problem with the Act is that, while it is very extensive in its operation, it is a toothless tiger when it comes to invoking penalties for non-compliance. Basically, if the Act is ignored there are no fines, penalties or retrospective procedures, unlike with the Building Act 1984 and the Town Planning Act 1990.

With the above in mind the writer feels the need to highlight some of the misunderstandings and mistakes that are being made in the implementation of the Act. That is to say that these same misunderstandings and mistakes were being made before, under the previous London Building Act. Sad to say that many surveyors who have been practising for years have not really considered the matter seriously enough and are finding that as the nation becomes more litigious, they are being questioned more on their actions and the Awards that they are producing.

The well-known case of *Gyle-Thompson and Others v Wall Street (Properties) Ltd [1974] 1 All ER 295* gives us a good basis on which to administer the Act properly. The judge in that case said, *".........It therefore seems to me important that the steps laid down by the Act should be scrupulously followed throughout, and short cuts are not desirable...............the approach of surveyors to those requirements ought not to be casual"*. It is with this in mind that the writer concurs with the view that if there are rules to follow then clearly they should be followed.

The purpose of this book is therefore not to castigate established surveyors but to assist their work, in order that the profession of practising party wall surveyors might be united in their interpretation of the Act, notwithstanding that, the advice given cannot be absolute, simply because of the grey areas and anomalies that exist within the Act itself.

The writer does not hold himself to be the ‘expert’ in the subject, but just one who has been practising since 1970 and who would therefore wish to pass on his views and considered opinions. He has taken legal advice on many matters concerning his opinions in this book, although the book is not intended to be an authoritive guide, but merely to highlight the misunderstandings and to aid surveyors on how best to administer and implement the Act.

It is said that the best writer is not one who writes to be understood but one who writes that he will not be misunderstood – a clear difference you will agree. Unfortunately the current Act has not been so written, hence the need for so many commentaries, but it is hoped that this book will be of good assistance to those who care to use it. The writer welcomes readers’ views and comments that could provoke the need for modifications to produce a second book.

At the end of each section are conclusions to summarise the points that have been raised in an attempt to clarify the matter.

The words Building Owner, Adjoining Owner, Notice and Award have been capitalised throughout this book simply to stress the subject matter and are not intended to carry any other significance.

References to the masculine are of course applicable to the feminine as is the singular and plural.

THE TITLE
Party Wall etc. Act 1996

It has been asked as to why the "etc" is in the title of the Act. Basically the answer is that the Act does not just apply to party walls themselves. A party wall is defined in section 20 of the Act and is quoted below.

The "etc" part covers party structures separating two owners, such as a floor between flats. By virtue of section 6, it also includes excavations within certain distances of the Adjoining Owner's structure.

Under the London Building Acts (Amendment) 1939 it was simply known as Part VI – Rights etc of building and adjoining owners.

To quote the preamble to the Act itself: – "*An Act to make provision in respect of party walls, and excavation and construction in proximity to certain buildings or structures; and for connected purposes.*"

As to Party Walls there are common law definitions:

1. A wall, which belongs entirely to one of the adjoining owners, but is subject to an easement or right in the other to have it maintained as a dividing wall between two adjoining properties.
2. A wall divided vertically into two strips, one belonging to each of the adjoining owners.
3. A wall divided vertically into halves, each half being subject to a cross easement in favour of the owner of the other half.

The statutory definition, as provided by section 20 of the Party Wall etc. Act 1996 is:

(a) *a wall which forms part of a building and stands on lands of different owners to a greater extent than the projection of any artificially formed support on which the wall rests; and*

(b) *so much of a wall not being a wall referred to in paragraph (a) above as separates buildings belonging to different owners.*

A wall may have part of its length or height forming a party wall, whilst the rest of that same wall exists only as an external wall.

The Party Wall etc. Act 1996 overrides any common law interpretations and rights, but only if the Act is invoked by the serving of a relevant Notice to undertake certain works specified by the Act.

SECTION 1
New building on line of junction

One

The line of junction that is mentioned here simply refers to what we might better know as the boundary. The line of junction is a line having no thickness but infinite length that separates lands having different ownerships.

It is clear from the heading that this section is applicable to new building on the line of junction, but already confusion has arisen between surveyors because of the phrase *'....built on at the line of junction..... '* Is the wall built on the line or at the line? This is merely a question of syntax and it is simply a wall built on the line at its junction or meeting point. This immediately tells us that it cannot be any distance away from the line but must be on the line.

Some have understood this to mean that the building must be astride the line in order for it to be 'on' the line. Some have considered that building a metre away from the line of junction is sufficient to meet the 'at' the line of junction, but surely this must be nonsense.

However, I believe that it is to be understood as building right up to and touching this imaginary line separating the two owners properties. Most surveyors would seem to concur with this view.

Two

What is important is that this section is applicable to not just new building work, but that the building must be on land not previously built on. That is to say a first floor side extension built above an existing ground floor extension, which is on the line of junction, would not apply under the Act because the land has already been built on at the line of junction.

However if the ground floor extension were to be demolished and the foundations grubbed up, then the land would be returned to its virgin condition and would be regarded as not built upon.

Despite arguments to the contrary, there is no time limit as to how long the land is not built upon and the action of demolishing and re-excavating could be within the same day. There are cases where this action could be argued so it may well be prudent to leave the land clean for one day and then start the new works.

Three

The question has been asked as to why the Building Owner would want to serve a Notice under section 1 in the case of a free standing building at the bottom of the garden when no one else is affected and it seemingly just simply attracts fees for party wall surveyors. In the light of this we must remember that the Act is meant to be an enabling Act.

For example if the building were set back from the line of junction for just say 5mm then section 1 of the Act would not be invoked. Assuming that there is no other structure on adjoining land within 3 metres whereby a section 6 Notice would have to be served.

I believe the simple answer is that by invoking Section 1 of the Act, the right of access under section 8 would kick into place and thereby the need to build overhand with all its difficulties would not be necessary. It is of course a question of costs weighed against the difficulties of building. It could prove very difficult if one is trying to build a two-storey structure near a line of junction with not having the right of access to the Adjoining Owner's land. The fees and possible compensation costs could well be worth it.

The other serious consideration is that to build a two-storey structure without scaffold would probably contravene health and safety regulations and therefore it could not be built, so it is extremely important in this case that access is obtained and this can only be given as a right under the Act.

There is of course divided opinion regarding the right of access, but the important thing to remember is, is it necessary and is it in pursuance of the Act.

Four – Section (1)(6)

A common error made with section (1)(6) is the perceived right to place foundations on the adjoining owner's land. The 1939 Act said the following:

'.........the building owner builds a wall on his own land he shall have the right at his own expense at any time after the expiration of six months from the service of the notice to place on land of the adjoining owner below the level of such land any projecting footings and foundation making compensation to the adjoining owner.........'

Whereas the 1996 Act says:

'Where the building owner builds a wall wholly on his own land................he shall have the right, at any time ....to place below the level of the land of the adjoining owner such projecting footings and foundations as are necessary for the construction of the wall'

The main difference is hidden in three words AS ARE NECESSARY. The Oxford dictionary definition of necessary is *essential, a requirement and needful*. Clearly these are strong words so if it can be proven that it is necessary, essential and needful to have your foundation projecting onto the Adjoining Owner's land then so be it. However I have to say that I do not know of any such case in construction where it is necessary for this to be so. It may well be that such an example springs to mind but it is difficult to imagine.

Foundations can easily be designed for eccentric loading so unless you want or agree to have your neighbour's foundation projecting onto your land then you do not have to allow it to happen.

The acceptance of projecting footings (brick corbelling) or foundations (concrete) will most certainly affect your property should you so wish to build against that line of junction.

One would of course certainly have the right to cut off any such projections as given in section 2(2)(g) but this could prove expensive, particularly if the projecting foundations are of the trench fill type. If the projecting footing or foundation is left in place then you would have to design around them with possible cantilevers, however small, but all at an additional unnecessary cost.

The projection of other parts of a building is dealt with under section 2 but it should be borne in mind that consideration should be given to such when building on the line of junction especially with regards to eaves and guttering overhang. Unless by agreement with the Adjoining Owner, this is not allowed and can be seen as simple trespass at common law.

Overhanging guttering can unwittingly become an easement by prescription, and removal at a later stage could prove difficult. Building on the line of junction should anticipate such potential overhanging and the detail should be modified, or could be asked to be modified by the Adjoining Owner. The modification could take the form of a parapet detail with a box gutter behind or a gutter that simply rests upon the top of the wall.

Five
There is the idea that if a Notice is served to build a wall astride the line of junction in accordance with section 1(2) and no consent is given the Building Owner must then build the wall wholly on his own land under section 1(5) and this would require a new Notice. Remembering that the purpose of the Notice is to give the intention to undertake certain works. Therefore as the works are different under section 1(2) and section 1(5) each require their own Notice.

This is of course wrong thinking because if consent under section 1(2) is not given then automatically section 1(4) comes into play whereby the Building Owner must build wholly on his own land and therefore the Notice served under section 1(2) covers the work under section 1(4).

Clearly one could also serve Notice under section 1(5) if a wall is to be built on one's own land in the first place, but the idea of serving two Notices is unnecessary and I believe will only serve to confuse the Adjoining Owner. Notwithstanding that the Act is quite clear as to what to do if consent to build across the boundary is not given.

Six
It has been asked as to why permission should be sought by a Building Owner from the Adjoining Owner to build a wall on his own land. Surely he can do what he wants on his own land. The answer of course is that under the Act he can't without serving a Notice of his intention. This is simply because his work might well affect the Adjoining Owner's land and the Adjoining Owner wants the right of protection as provided by the Act.

However, all is not lost because it is generally to the advantage of the Building Owner insofar that section 8 of the Act would be invoked, which deals with the right of access. It would be more difficult to build a wall overhand on the boundary than being simply able to stand and face the work, which ultimately is usually more aesthetically pleasing to the Adjoining Owner who would be the one viewing the finished work.

Seven

A question arises as to when construction has occurred on one side of the line of junction, and whether this would preclude the serving of a section 1 Notice by the neighbouring owner on the assumption that the line of junction is not built on.

There is the belief that it does so preclude and the line of junction can be said not to be built upon on one side of the line. If a Notice could not be served in the case where the line of junction has been built upon on one side alone, then this would not be equitable. The loss of being able to serve a Notice would therefore preclude the Building Owner from undertaking his work properly, or maybe even at all! However this has not been tested at law and it would be for surveyors to agree or disagree this one.

If, however the line of junction is say 10 metres long and it has been built on for only 5 metres of its length, this does mean that most certainly the rest of the line of junction is still available for use.

Eight

It has been suggested by some that a Notice served under sections 1(5) and (6) gives the right to remove a garden boundary wall or fence owned by the Adjoining Owner in order to construct a building. I do not believe that this is correct and logic tells us that this cannot be so, i.e. to remove another person's property in order to accommodate your own. I cannot imagine any judge upholding such a request.

However those same would say that section 8(1) would be invoked and the owner has the right *'in pursuance of the Act'* to *'remove any furniture or fittings or take any other action necessary for that purpose'*. This is taken to mean that a garden wall or fence could be removed, but I find that this must be a matter of degree and I can see that a third surveyor could well be called upon to determine such an issue.

Nevertheless I disagree with such arguments, for the same reason given previously that a foundation may be placed upon the adjoining owner's land by virtue of section 1(6) as that word 'necessary' has been used again. So unless it is 'necessary' to remove such fences and walls to build, then there is no right to do so.

Nine

There is an intriguing anomaly about serving a Notice under section 1 because the Act only states what happens when consent is given and not what happens when the Notice is dissented to. This section of the Act never speaks about a *'deemed dispute'*, that is, when a Notice has run its 14 days without reply as opposed to sections 2 and 6. It has always been taken to create a 'deemed dispute' from the general spirit and tenet of the Act in other parts, and this anomaly has merely been put down to bad drafting.

What this section of the Act does say however, is that *'where a dispute arises under this section between the building owner and any adjoining owner or occupier it is to be determined in accordance with section 10'*. This means that surveyors are appointed to resolve the matter.

Ten

It should be carefully noted that the position of the legal boundary must be determined or agreed between the owners and is not a matter within the remit of appointed party wall surveyors, albeit that they can offer much help, and indeed the guidance of the surveyors might well be sought by the owners. In the event of a dispute however, party wall surveyors cannot determine the boundary in their capacity as such.

Conclusions

1. A building must be on the line of junction for this section to be invoked.
2. The land must not have been built upon at this junction.
3. Right of access is due because the works would be in pursuance of the Act.
4. The right to place foundations on an Adjoining Owner's land has to be necessary.
5. Only one Notice is required to be served.
6. The Adjoining Owner has rights of protection.
7. A Notice to build can be applicable to both sides of the line of junction where no structure exists.
8. Removal of property is not permitted, without consent, save as to it be proven necessary.
9. Surveyors settle any dispute between the parties.
10. The position of the legal boundary is not for party wall surveyors to determine.

SECTION 2
Repair etc. of party wall: rights of owner.

There are thirteen cases listed where the Building Owner has certain rights to undertake works affecting a party wall, and each has its own anomaly.

One - Section 2(2)(a)
This subsection deals with the party wall itself in that it can be underpinned, thickened or raised, and each of these have presented misunderstandings.

UNDERPINNED – this could be in a situation when a new basement is formed and the Building Owner wishes to underpin the wall to the same width as the existing strip foundation or corbelled brickwork.

The answer is twofold in that he cannot underpin to the same width because in forming a basement the Building Owner is not actually underpinning but building a wall and retaining the adjacent ground. The wall is only the thickness of the upper part. The corbelled brickwork is known as footing and is built to spread the load of the wall above. The foundation is the strip of concrete, however wide and thick, so positioned to support the wall above.

It therefore follows that as the Act refers to the wall then only the thickness of the wall must be taken. The wall stops at the top of the brick footing or concrete strip.

The second reason for not allowing the full width to be taken is that it could be prejudicial to the Adjoining Owner if he wished to create a basement and space/area would be lost by the encroachment of the concrete foundation/wall.

It may well be that the wall will have to be redesigned by the engineer to be slimmer, which probably means more reinforcement. The encroachment of the foundation/wall could be seen as trespass at common law and consideration should be given to this aspect also.

The truth is that when a basement is formed below existing party walls this is not ‘underpinning’ as in foundation work, but is in fact the formation of walls. If such walls, even though they may be of concrete containing reinforcement for stability in that they are retaining land on either side, the walls are not to be considered as ‘special foundations’ but are simply reinforced walls.

It follows then, that if the Adjoining Owner subsequently excavates to form a basement for himself, the reinforcement in the concrete wall, which was once placed there in order to retain, will now be redundant. It is with this in mind that a conclusion to keep the thickness of this wall to be the same as that above it should be made. In other words the new basement wall should not extend on to the Adjoining Owners land save by agreement between the owners.

Two

THICKENED – This is meant to be what it says to thicken a wall, which for example could be the increasing of a half brick party wall to a one brick party wall, as would be needed for structural reasons. However, the question has arisen as to whether any thickening at all would be notifiable work under this section for such work as the rendering of a wall.

I believe that such work is not notifiable because rendering and plaster is regarded as a finish, and not an integral part of the wall. It is not required for structural integrity. The question arose because render was to be applied as a waterproofing agent and thus giving protection to the wall. However the wall could be weather protected with silicone treatment and therefore, if we are to consider a finish as thickening the party wall, then we should also consider even paintwork which would of course be nonsense.

Three

RAISED – Again this simply means what it says and would occur in such cases as raising a party wall between two single storey extensions to form first floor extensions. It could also occur in the case of infilling the party wall void between two roofs in a terrace situation which can still be quite commonly seen. The wall in this case is being raised from the ceiling level to the finished roof level.

Another example would be the raising of a parapet wall to form the flank wall of a dormer/mansard roof construction. One needs to be carefully aware in such a scenario, because many Local Authorities would consider that such action requires planning permission, insofar that the Building Owner is not building on his own land. A Planning Notice would therefore have to be served on the owner of the land, together with a Certificate of Ownership being sent to the Local Authority. This is all simply because the boundary has been breached, and occurs irrespective of any permitted development rights.

However there is at least one such Town Planning appeal that I know of whereby the inspector has allowed the appeal as he considered that such a wall came within the curtilage of the dwelling, and granted a Certificate of Lawful Development.

This sort of decision is not a party wall matter but clearly affects a Building Owner's proposal and thus what is notifiable work. A similar argument does of course ensue on building any wall across the boundary and Town Planning rules should be considered.

Four – Section 2(2)(c)
A partition that *'does not conform with statutory requirements'* does not mean that a party wall needs to be re-built to the latest Building Regulation standards every time the regulations change. The majority of party walls would not comply with the current standards for thermal and sound.

The sort of partition that might be considered to be re-built would be the one that has been erected to separate two flats in a converted house. Such a partition might only be just one leaf of timber studwork covered with plasterboard.

The Act actually covers this situation in section 2(8) when it says that if such a wall complied with the regulations before the Act was applicable on the 1st July 1996 then it is *'deemed to conform with statutory requirements'*. Walls that are built after this date need to comply with the current Building Regulations, but however, would not need to be continually changed as new regulations subsequently arise.

Five – Section 2(2)(f)
This is an interesting one – *'to cut into a party structure (which may be or include the purpose of inserting a damp proof course);*

It is rare these days to cut into a party wall to insert a damp proof course for it usually done by drilling and chemical injecting, so where would this stand within the Act?

In strict legal definition terms, 'cutting' may not be the same as 'drilling' and 'inserting' is not the same as 'injecting'. Nevertheless they would fall within the spirit of the Act rather than the letter of the Act and most would consider this to be so.

However, unfortunately, chemical injection as a method of DPC renewal brings with it many problems. If there is a plaster finish to the internal face of the wall then this is to be removed, usually to a height of at least one metre and re-render and finish etc. is required.

If this method is used on a party wall then the plaster finish to the Adjoining Owner's side must be replaced and therefore the Building Owner would have to replace this. Notwithstanding that the costs should be shared between the parties, an insertion of a physical DPC would be cheaper insofar that damage would be below the skirting board level, and repairs of any kind would not be required to the Adjoining Owner's property (as well as the Building Owner's property).

There is also the argument against chemical DPC injection in that there is no control over where the fluid goes and could be considered an act of trespass. It cannot be removed, whereas a physical DPC can be.

An agreement to undertake such works in whatever form is required between owners, via the appointed surveyors. The Adjoining Owner cannot disagree in principle to the works, as the Building Owner has a right to take any action that does not allow his property to fall into disrepair and dilapidation. This section would be covered by section 2(2)(b) that allows *'to make good, repair....a party structure....where such work is necessary on account of defect or want of repair of the structure of the wall'*. The action is covered by case law such as *Bond v Nottingham Corporation [1940] Ch 429.*

This is also the section that is used for the insertion of steel beams into a party wall for loft conversions and through lounges etc. It is therefore important to establish the thickness of the party wall as many surveyors have been caught out and discovered that the wall is only a half brick thick. In this situation the structural stability of the wall must be assessed to determine if the wall needs to be stiffened or underpinned. The wall may be a timber frame construction and the appropriate detail must be ascertained.

It is usually taken as understood that the Building Owner will not use the whole thickness of the party wall in which to take bearing for his steel beams, but technically as the whole of the party wall is available for use, it is usually best to state in the award that the steel beam be inserted only up to half the thickness of the wall. This ensures that the beam is not projecting into the Adjoining Owner's loft for example, and that fire integrity is maintained between the dwellings.

The application of this section is exercisable subject to making good all damages occasioned by the work as defined in subsection 2(5).

Six – Section 2(2)(g)
This is about cutting away projections from a boundary wall in order to build a wall against it. This is a tricky one and care must be taken as many problems can arise, particularly as there would seem to be a contradiction between this part of section 2 and section 9 that deals with easements.

There is an apparent contradiction here because on the one hand the Act is saying that you can cut off any projection and then on the other hand saying that the Act does not over-ride any easements, which may of course entail a projection.

Firstly to cut away a foundation could be more costly than building new work around it. The foundation could be a special foundation containing reinforcement. The foundation could be deep trench fill so a lot of cutting away would be required.

Fortunately there is case law on the matter, for example *Arena Property Services Ltd v Europa 2000 Ltd (Court of Appeal, 2003).*

This concerned the removal of a soil and vent pipe protruding onto the adjoining property, thus affecting the proposed works of the Building Owner.

As is in that case the simple answer is that an easement is not an easement unless there is a benefit to the dominant owner. So for example one could not remove overhanging guttering because it would clearly be in use every time it rained and thus fulfilling its purpose.

One could think of a host of other such examples, such as embellishments to walls that project beyond the boundary, which in accordance with the Act could be cut off, as this could not be said to be of any benefit to the dominant owner. However this could well be subject to other rights such as would be contained within the laws for Listed Buildings.

There is the matter of the Act saying that you can cut off projecting flues, and would you say that this is an easement? It might well be proven to be so if the flue was now serving a gas appliance or back boiler. What about ordinary balanced flues from gas boilers? One would be very foolish to just cut these off.

The answer has to be one of common sense and agreement between owners and surveyors. Despite a soil and vent pipe being an easement it could be removed by agreement. If the Building Owner is prepared to re-site the pipework or a boiler flue at his own expense for the Adjoining Owner, an agreement could well be reached, but if there is no agreement then the Building Owner would simply have to build around the obstacle.

It has to be borne in mind that if the boiler flue for example has been so positioned comparatively recently and discharges over his neighbour's land, then it could be requested that it be removed at the sole cost of the Adjoining Owner on the simple grounds of trespass.

If however the boiler flue has been in that position for some twenty years or more, then prescriptive rights may have arisen, establishing an easement. Further comments have been given on this matter under section 9.

The application of this section is exercisable subject to making good all damages occasioned by the work as defined in subsection 2(5).

Seven – Section 2(2)(h)

One should be careful in using this subsection, insofar that whilst it would appear that a right is given to cut off any projections in order to enable a vertical wall to be erected or raised against the wall, any easements must be considered that might be in place. The building could of course be a 'listed building' whereby other laws are applicable and would override this subsection of the Act.

The application of this section is exercisable subject to making good all damages occasioned by the work as defined in subsection 2(5).

Eight – Section 2(2)(j)

It seems absurd to have to serve a Notice to claim the right to insert a weatherproof flashing into an Adjoining Owner's wall, but very often it has proven necessary to do simply because the Adjoining Owner has refused to co-operate under normal neighbourly agreements. Such a flashing would be of benefit to both parties to seal off the affects of weather and the possible ingress of water, which could cause damage to both properties.

Nine – Section 2(2)(k)

A simple sub-section but a 'catch all' and should invariably be used in conjunction with other party wall works where connection is made to the adjoining premises. It often allows for the unforeseen and incidentals of the work without the need to have to serve further awards.

Ten – Section 2(2)(m)

This sub-section gives a right *' to reduce, or to demolish and rebuild, a party wall or party fence wall to –*

> *(i) a height of not less than two metres........................'*

The question that has been raised is, what if the wall separates land of two different levels? Indeed quite a common phenomena, at what height should the dimension of two metres be taken, from the low side or the high side? I think the answer has to be from the high side as the lower portion is acting as a retaining wall and could in fact be even two metres in height itself.

The fact of the matter is that the sub-section is somewhat superfluous because the adjoining owner could serve a counter notice under section 4 to have the wall built to its original height, as duly mentioned in section 11(7). The chances are of course is that if the wall is a retaining wall it is not a party wall but in the ownership of the land being retained, as clearly when the wall was originally built it was so positioned to retain the land of the 'builder' and not built by an adjoining owner to hold back another persons land. However there will naturally be cases of the opposite situation but most evidence would point logically to the former.

Eleven – Section 2(2)(n)

Lateral thinking is required for this sub-section because a wall can be exposed on its top side and not just on its face. Such is the case of re-roofing over a party wall or the removal of parapet coping stones, which after all have only been placed there for weathering purposes. The general consensus is that a party wall that is exposed on its face is notifiable work, but one that is exposed momentarily on its top side is not.

It has never been considered a problem to demolish a structure against a party wall thus exposing the wall to the elements. Under such circumstances the Building Owner is required to make the exposed wall weather tight.

The Act, however, is unclear as to who should maintain this weatherproofed wall. If the wall were to be left permanently exposed, then the usual method would be to render the wall. The possibility arises however that that the Adjoining Owner could ask for better protection such as an additional skin of brickwork. This is on the reasoning that the assumption will be that the Adjoining Owner will be left to maintain the wall in the future, and why should he therefore be forced to have this expense?

If the wall is not abandoned as a party wall, then there is an argument that the Building Owner should still be responsible for its upkeep. Under such circumstances the Adjoining Owner would have the right under section 2(2)(b) to maintain the wall and so request a due proportion of payment from the Building Owner.

If the wall is abandoned as a party wall and deemed to be an external wall to the Adjoining Owner's property then he would only have the right of access for maintenance and repair under the *Access to Neighbouring Land Act 1992,* with the possibility that he might have to pay compensation to the neighbour. This is why the Adjoining Owner could request a full brick veneer to be added to avoid this future cost and inconvenience that he might suffer.

The Act does not make any provision when such weatherproofing of an exposed party wall takes place as to who should be responsible to maintain it in perpetuity. Equally, the Act makes no provision as to what happens to the wall in future; hence the surveyors should make the determination by Award.

Conclusions

1. A new basement wall should be to the same thickness of the existing wall unless differs by agreement between owners, and is not underpinning.
2. Finishes are not part of a wall with regards to thickening.
3. Raising a party wall may be permitted outside of any Town Planning requirement.
4. The wall does not need to be constantly upgraded to meet the latest Building Regulation standard.
5. Check the DPC method implications and the thickness of the wall.
6. Easements should be considered with regards to cutting away projections.
7. Check other rights before cutting away projections.
8. Insertion of flashings against a neighbour's wall is of right.
9. Allow for the unforeseen.
10. A reduced wall can be re-built to its former height.
11. Exposing a party wall could be expensive.

SECTION 3
Party structure notices

One

The Act says in section 3(1*), 'Before exercising any right conferred on him by section 2 a building owner shall serve on any adjoining owner a notice (in this Act referred to as a "party structure notice")'*

It can be seen that there is a reference to section 2 works that a Notice should be served in the manner set out. Strictly speaking section 3(1) does not therefore relate to works covered by section 1 or section 6, albeit that these sections also require a Notice. In those sections the format of the Notice is not described. However it has always been taken that any Notice should be in the format laid down in section 3(1).

One of the most crucial things to understand about the Act is that unless a notice is served then the Act is not invoked. What therefore is a Notice? A Notice is simply conveying information about the intention to undertake certain works by the Building Owner to the Adjoining Owner that could have an affect upon the Adjoining Owner's property.

Perhaps an easy way to understand the principle of a Notice invoking the Act would be if you were parked illegally, whilst you would be due to pay a fine it cannot be paid unless a ticket is deemed served on you, thus invoking parts of the relevant road traffic regulations. Therefore there is no mechanism to pay a fine without being served a valid ticket. The case of *Emms v Polya [1973] 227 EG 1659* said that failure to serve a Notice was not in itself unlawful.

Although a legal obligation is laid upon the Building Owner to serve the Notice, the Notice does not have to be drawn up by any legal expert, and can simply be hand written, and providing it contains the required information, becomes a legal document. There are of course standard forms that have been produced by various professional bodies to help with the matter and clearly the Building Owner may wish to avail himself of these.

Very often a Notice is served on behalf of the Building Owner by other parties such as a solicitor or much more commonly a surveyor. At this point it must be made clear that a surveyor who serves such a Notice is not an appointed party wall surveyor, but merely as an agent acting on behalf of the Building Owner.

It is very important to note that one cannot be appointed as a party wall surveyor until a Notice has been served and dissented to, which the following cases confirm; *Adams v Marylebone Borough Council 1907 2KB 822, Emms v Polya [1963] 227 EG 1659, Leadbetter v Marylebone Corporation [1904] 2 KB 893, London & Manchester Assurance Company Ltd v O & H Construction Ltd [1989] 2 EGLR 185,* and *Louis v Sadiq [1997] 1 EGLR 136.*

The simple reason being is that it could well be, and often is, that consent is given to the works. Therefore a party wall surveyor is not required and cannot be appointed. The purpose of a surveyor is to resolve a dispute between the parties and if there is no dispute, it follows that there is nothing to resolve, and surveyors need not be appointed.

The content of the Notice is set out in section 3(1) and gives exactly what is required. If it is followed it will be legal and valid, if not it could be invalid, and everything that follows will also be invalid – clearly this could be quite catastrophic to all parties concerned.

It is very important to note that if a Notice is served there is nothing to prevent any agreement being made between the two owners. What is important in such a case is that the agreement should be in writing.

This ensures that no change of mind occurs when the Building Owner is half way through the project, albeit the Adjoining Owner would be hard pressed to complain, as he would have been seen to have acquiesced to the works.

What is not clear in the Act is that whilst the Adjoining Owner can give consent to the works in a Notice, it does not automatically mean that all rights have been signed away, and any future dispute during the same works between the Building Owner and the Adjoining Owner must be resolved at common law.

Most surveyors would claim this route, but the Act does not say that rights are automatically removed, indeed section 10(1) says that *'Where a dispute arises or is deemed to have arisen between a building owner and an adjoining owner in respect of any matter connected with any work to which this Act relates either –*

> *(a) both parties shall concur in the appointment of one surveyor...............or*
> *(b) each party shall appoint a surveyor ....'*

So following this, a dispute between the owners could arise after consent is given that needs to be resolved. In which case some would consider that Section 10 (1) would 'kick in'.

A Notice must state three things for it to be valid, and these are given in section 3(1)

> The name and address of the Building Owner.
> The nature and particulars of the work.
> The date on which the proposed work will begin.

Many surveyors make the mistake in declaring a Notice invalid if it does not contain the various clauses set out in section 2(2)(a) to (n). This is of course incorrect for one must remember that it is the duty of the Building Owner to serve the Notice and such a person will probably not have any real building knowledge let alone understand the Act. Providing the general principles of what the works are contemplated are stated, then that is sufficient for the Notice to be compliant. In fact it is considered that Notices ought to be construed fairly liberally, as the Act makes no provision for amending a Notice.

Two

At this point it should be noted that a Notice is a legal document and as such all legal documents and papers are *served.* They are not issued, sent, delivered or worst of all, published. Surveyors commonly use this word term of *publishing* as though one has to use special jargon. Publication would be a broadcast to all and sundry, whereas service is particular upon the owners and this is a private matter between the owners.

It may interest you to know that the mention of *serve, serves* and *services* occurs no less than 51 times in the Act and no other word is used. The words *delivering*, *delivered* and *sending* are used in the Act (section 15), but only to describe the method by which the Notice or Award is to be served. This is discussed later in the book under SECTION 15.

In the London Building Act part VI, the forerunner of the 1996 Act, the word *publish* did not appear either - but the words *serve, serves, service and served* occurred 21 times.

Three

As a Notice is a legal document it needs to be carefully drafted, albeit that the statement concerning the works should be liberal to allow for the unforeseen. One of the common mistakes made by surveyors is getting the rightful owners wrongly named in the Notice and in the Award. Such a mistake could invalidate the Notice and subsequent Award.

When serving a Notice the correct Building Owner(s) must be stated. For example, if the owners are Mr Brown and Mr Green and then the Notice is served and merely says Mr Brown, the Notice would be deemed invalid. Perhaps less obvious would be if Mr and Mrs Brown are the owners but only Mr Brown is named as the owner. There is of course case law highlighting such mistakes such as *Lehman v Herman [1993] 1 EGLR 172* whereby such a mistake occurred and all that followed was deemed to be invalid.

The situation is different for Adjoining Owners who in the Notice do not need to be named. The Act states that a Notice is to be served on *'any adjoining owner'*. It is therefore sufficient to serve a Notice on *'the owners'*, or one of the co-owners, such as the husband or the wife, but all separate owners, such as leaseholders and freeholders of the same property are entitled to receive a Notice.

The case of *Crosby v Alhambra Company Ltd [1907] 1 Ch 295* highlighted the definition of owner and adjoining owner and that both freeholder and leaseholder were entitled to a Notice. The judge defined that whilst all owners with a separate legal estate were entitled to a Notice, it was only necessary to serve Notice on a single co-owner. In other words husband and wife would constitute being co-owners, and as such only one of them needs to receive a Notice.

Having said that, it is generally very easy now to find out the correct name(s) of the owner(s) by logging on to the Land Registry website and for a small fee, the names of the registered owners can rightfully be established. Of course not all properties are registered but most are, and, in the next few years they are likely to be. Nevertheless beware of taking the Land Registry details for granted, as sometimes it may well have not been updated by the inclusion through marriage or change of maiden name to married name.

In the case of Limited Companies as Owners, the names of the directors can be checked at Companies House free of charge.

Four
One of the common mistakes made by surveyors is the misunderstanding of section 3 (2)(b) and also again in section 6(8)(a) that says:

> '(b) cease to have effect if the work to which it relates:-
>
> (i) has not begun within the period of twelve months beginning with the day on which the notice is served; and
>
> (ii) is not prosecuted with due diligence.'

Many surveyors understand that the work can cease to have effect if there is a pause in the work and cite (ii) above, for this is a linked stand-alone sub-section as it is linked with (i) above in the use of a semicolon i.e. a semicolon is greater than a comma but less than a full stop. Therefore both can be operative separately and not to be read with just the conjunctive AND. If the work has indeed started within the twelve months then (ii) can be invoked alone under the Act.

There are those that would read the linking of the two sub-sections with the AND, and therefore assume that they must be read together. If this were so then sub-section (ii) could never be invoked because sub-section (i) would have been complied with in that the works have commenced within the twelve-month period.

There could be legitimate circumstances whereby the twelve month period has expired, but for good reason the work has been prosecuted with due diligence. Equally the work could have commenced within the twelve month Notice period but for legitimate reasons the works has had to be halted for a period of time.

It must be noted that the twelve-month time limit does not apply while a dispute is still being resolved pursuant to section 10 of the Act: *Leadbetter v Marylebone Corporation (No.2) [1905] 2KB 893.* This case, although old, actually lends weight to its authority.
The case does not seek to extend the validity of the Notices. It merely states that the time period specified for Notices is relevant only when there is no dispute between the owners. It is in fact very much a commonsense decision.

The cited case is in fact supported further by the case *Theresa Filippi v Beatrice Jakob [2001] EWCA Civ 827.* covering the 1939 Act in which 'Leadbetter' case was referred to.

Once an Award is served, the position is similar to as if the work is consented to under a Notice, except that now the issues in dispute have been resolved. It is therefore sensible, and if anything, a safeguard for the Adjoining Owner, to insert a clause stating that the works should commence one year from the date of the Award. Albeit that such a clause is not necessary to validate an Award.

Five

The period of time mentioned as being one and two months means that a month is one calendar month and not 28 days, and commences from the date of the Notice as given in section 1(5) (a), section 1(6)(a), section 3(2) (a) and section 6(5).

Conclusions

1. The Act cannot be invoked without a valid Notice being served.
2. Notices and legal documents are 'served'.
3. The correct Building Owner must be named in the Notice.
4. Work can be intermittent.
5. The time period is given as calendar months.

SECTION 4
Counter notices.

This section deals with Notices that can be served by the Adjoining Owner. It comes about when the Adjoining Owner receives a Party Structure Notice from the Building Owner and is not in agreement with the proposals.

The Adjoining Owner has the right to serve on the Building Owner a Counter Notice giving details of his own requirements for works that should be undertaken at the same time.

The Adjoining Owner can ask under section 4(1)(a) for such things as the building of *chimney copings, breasts, jambs or flues, or such piers or recesses or other like works, as may reasonably be required for the convenience of the adjoining owner;*

He can also request under section 4(1)(b) that, having consented to special foundations being built on his land, that they;

(i) be placed at a specified greater depth than that proposed by the building owner: or

(ii) be constructed of sufficient strength to bear the load to be carried by columns of any intended building of the adjoining owner,

or both.

In such cases the request is dealt with by the surveyor(s) in the preparation of the Award, to which they have powers under section 10 (10) and (12). Nevertheless it is for the Adjoining Owner to serve the Notice and not the surveyor unless of course he has been empowered by written authority to do so.

Like any other Notice the Counter Notice must specify the works required and in addition the Notice should be accompanied with plans and details outlining the requirements.

The Counter Notice must be served within one month of having received the Party Structure Notice.

It is the requirement of the Building Owner to comply with the Counter Notice unless any or all of the threefold test applies which are as stated in section 4(3)

(a) be injurious to him;
(b) cause unnecessary inconvenience to him; or
(c) cause unnecessary delay in the execution of the works pursuant to the party structure notice.

Conclusions

1. Counter Notice must be served within one month of receiving Party Structure Notice.

2. Counter Notice must be complied with unless the reasons for non- compliance is given as stated in the Act.

SECTION 5
Disputes arising under sections 3 and 4.

This small section deals with the time period in which response to any Notice is served. The Act says that unless consent is given within fourteen days of receiving the Notice then a dispute is deemed to have arisen.

A dispute, deemed or otherwise, is then dealt with under section 10. The time period of fourteen days is good in that it stops owners trying to thwart the proceedings of the Act by trying to cause delays.

There is however nothing in the Act to say that consent to any matter cannot be given after fourteen days. The purpose of section 10 is to resolve disputes, but if during the process of resolving the dispute the owners agree upon a matter, then there is clearly no dispute to resolve and the surveyor(s) should not continue with their work.

It is a misunderstanding that the whole process of serving an Award must be carried through to a conclusion just because section 10 has been invoked by dissention, or deemed dissention, to a Notice.

Conclusions

1. If consent is not given to any Notice within fourteen days then section 10 of the Act must be invoked.
2. Consent between the parties can be made at any time even after the expiry of the fourteen days.

SECTION 6
Adjacent excavation and construction

Section 6 only deals with excavation within three metres and six metres of an Adjoining Owner's structure. The measurement for a three metre Notice is taken from the face of the proposed excavation to the face of the structure, and is measured horizontally, which is likely to be the projecting footing or foundation of the Adjoining Owner's property. The measurement is not taken between the faces of the walls, or from the face of the wall to the boundary. The boundary is in fact irrelevant to this section of the Act.

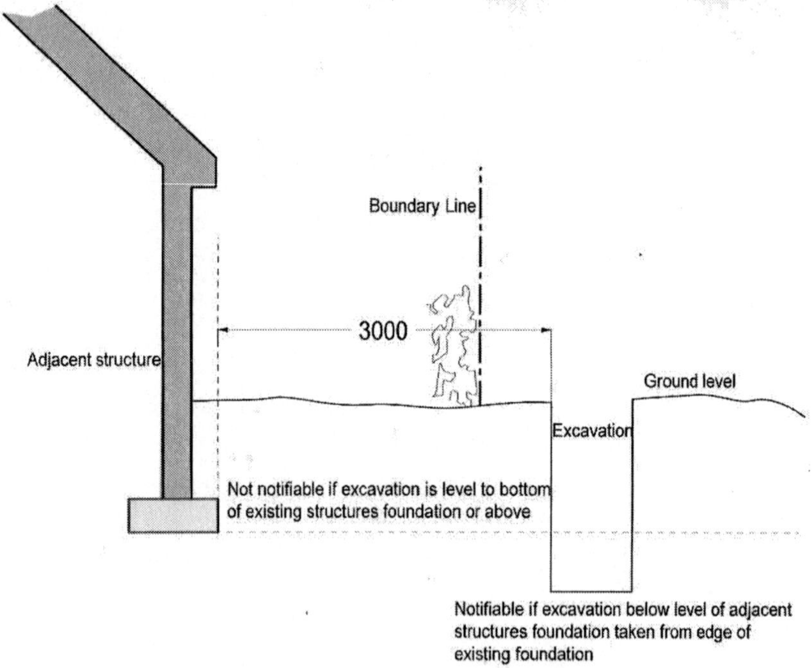

3 METRE TO EXCAVATION

The measurement taken for a six metre Notice is taken from the face of the excavation to the face of the structure, again measured horizontally, but the line of 45 degrees is taken from the face of the wall of the Adjoining Owner's property measured from the bottom of the foundation.

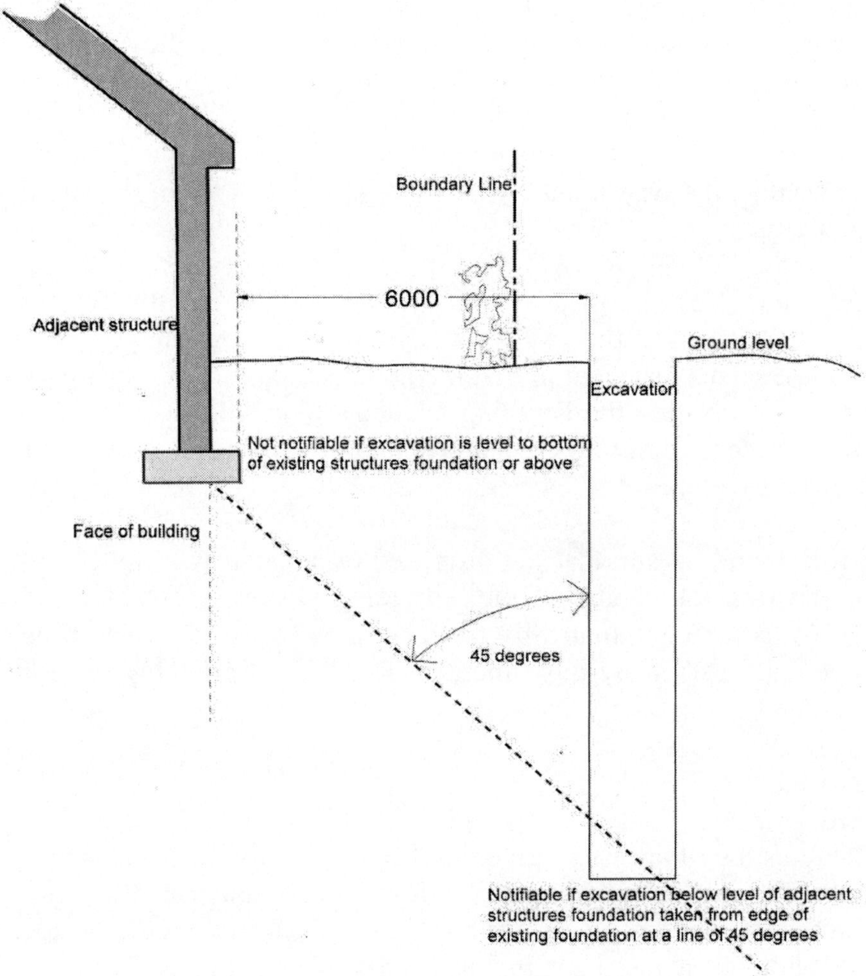

6 METRE TO EXCAVATION

One

Both the three metre and the six metre criteria have two parts, and both must be relevant for a Notice to be served, and they are:

a) the distance between the structure and the excavation.
b) the depth of the excavation.

This section is only operative when both parts are applied. If the distance is less, but the excavation is likely to be shallower than the Adjoining Owner's building, then the Act does not apply.

Similarly, if the distance is greater but the excavation is likely to be deeper than the Adjoining Owner's building, then again the Act does not apply.

The distance can be measured at ground level, even allowing for an intelligent guess at the likely foundation projection. However, if the distance seems marginal in that the depth is not known, together with the possibility that the Building Owner's foundation is not likely to go below the Adjoining Owner's foundation, then the general rule is to serve a Notice.

If it is found on site that the proposed excavation will not be below the existing foundation of the Adjoining Owner's foundation, then the Notice will automatically be invalid, and no further action needs to be taken, albeit by this time an Award could probably have been served.

Two

The question of access as given in the Act by section 8 seems to arise frequently with the application of section 6. Basically the Notice states the proposed works for any section, but this section 6 speaks only about excavation and that is all that should be considered.

Many surveyors become over enthusiastic in using section 6 and include all sorts of clauses in their Awards. Many have actually refused to relinquish their position in the project until all the works have been completed. This is of course fundamentally wrong, particularly as this section only deals with excavation.

Once the excavation has been finished the Award is completed, and there is no more for the surveyor(s) to do apart from check the schedule of condition and ensure that all is well for the appointing owner and the property.

Confusion has arisen because the section is entitled *Adjacent excavation and construction.* The understanding should be that the excavation is for the erection of a building or structure, and not a right to erect a building. The underlying harm that this section aims to deal with is that caused by excavation

Three

It is quite common to read in some Awards references to scaffolding and the dismantling thereof, but not the erection. It is difficult to perceive as to why scaffolding is required for excavation works, though there may be some cases that possibly make it necessary.

It should be noted that the right given under section 8 relating to access is to be only deployed in the works of excavation, and not for the rest of the building works. This tells us that scaffolding or any form of trespass cannot be allowed after completion of the excavation in relation to a section 6 Notice.

The reason being is that the right of access prescribed by section 8(1) is granted only for works in pursuance of the Act. Excavation, and anything related thereto, under section 6 is the only part that would be in pursuance of the Act. Therefore works not contained in any Notice would not be in pursuance of the Act.

It could of course well be that the Adjoining Owner will allow his land to support scaffolding for the Building Owner's work, but it may be necessary to offer compensation and indeed draw up a scaffold licence. A model licence can be found in the appendix of this book.

If excavations are very close to the line of junction then section 7 could be invoked.

Four

Section 6 of the Act has brought about long discussions with regards to the use of piling, insofar as the question is raised - is piling excavation?

Piling is a method that clearly has not been considered by those drafting this Act, which is unfortunate as an opportunity has been missed to update the section to cover piling and would have answered all the queries that have been raised.

The Act mentions excavation and the problem lies in whether piling is excavation. It certainly is when augured piles are used because earth is removed out of the top of the drill and is of course excavated material. The matter is different when driven piles are used as, strictly in accordance with the Act, nothing is excavated.

Nevertheless the majority of surveyors have considered that the displacement of soil amounts to the same thing. This is questionable but until it is tested at law most surveyors will continue to serve Notices for driven pile work, as damage certainly could occur with this type of piling and prudence is the better part.

The other matter about piling is that it is also designed for friction fit for its strength, rather than relying on bottom ground bearing as in conventional foundation design.

Conclusions

1. Both distance and depth together are necessary for this Section.
2. No work other than excavation is to be covered by this Section.
3. Right of access is limited to such work.
4. Piling should be considered as notifiable work under this section.

SECTION 7
Compensation etc.

One – Section 7 (1)
'Unnecessary' inconvenience caused to an Adjoining Owner or Occupier clearly shall not be made, but the wording does therefore by implication hold out that 'necessary' inconvenience is allowable, which quite often must be so, by the very nature of some of the works.

It should be noted that this section covers the effect that the works may have upon occupiers as well as owners, and so therefore would include tenants and lodgers if they were so inconvenienced by the works.

Two – Section 7(2)
The Act says that the Building Owner shall compensate for any loss or damage so caused because of the works, which is to be naturally expected in any circumstances.

Compensation should only be paid as is reasonably lost by virtue of the inconvenience of the works. It has to be measurable in financial terms relating to loss and damage. I believe that the damage mentioned is intended to be taken as physical rather than damages relating to loss of light etc, which effectively is an easement and not part of the Act. Compensation is unlikely to be given for just mere loss of garden enjoyment and the like under this section, but it is thought that 'loss' would include such things as having an affect upon ones business.

I had such a case whereby the Adjoining Owner was a writer and needed a very quiet environment in which to work. Compensation was awarded insofar that he had to move to an alternative accommodation during the noisy part of the works.

Alternatively, perhaps loss does cover everything, but for damages to be awarded it must pass the de minimus threshold

Three – Section 7(3)

Some difficulty has arisen with this sub-section in that the Building Owner shall at his own expense make necessary proper protection to the Adjoining Owner's property when it is laid open due to the Building Owner's works.

Such an example would be in the case of section 6 when excavations are close to the boundary and protection to the adjoining property must be given by means of hoarding or screens. These would have to be placed on the Adjoining Owner's land, and this is very often where the disagreement lies in that the Adjoining Owner does not want such loss of use of his land.

The Act gives a right under section 8 for access for any works that are in pursuance of the Act and therefore the Building Owner can invoke such. In addition this section 7 makes a right to be given to protect the Adjoining Owner's land against the works. This could be enforceable from a Health and Safety aspect by the Building Owner as the ground could be laid open to deep excavations, or maybe the adjoining lands are at different levels.

A cross reference can be made between this part and section 11(6) where *'a fair allowance for in respect of disturbance and inconvenience shall be paid',* but it is interesting to note that section 11(6) only makes reference to works carried out under section 2(2)(e). It does not seem unreasonable to suppose that disturbance and inconvenience is likely to also take place for works under section 7(3).

Conclusions.

1. Inconvenience could include all those resident at the premises.
2. Compensation could be awarded if the works affected a business.
3. Consideration needs to be given for site protection.

SECTION 8
Rights of entry

This is a section that has been misused and abused by many surveyors. The section clearly tells us:

Who may enter	section 8(1)
When they may enter	section 8(1)(2)
Why they may enter	section 8(1)

One

As to 'WHO' may (not shall) enter is defined as *'A Building Owner, his servants, agents and workmen', which* covers virtually anyone from the Building Owner's side associated with the work.

'WHEN' they may (not shall) enter could be debatable because *'usual working hours'* would be taken to mean times as laid down by most Local Authorities for contractual working.

The Control of Pollution Act 1974 based upon the Code of Practice BS: 5228 1984 'Noise Control on Construction and Open Sites', gives Local Authority environmental health departments special powers to control noise on construction and demolition sites. Working hours can be restricted and conditions enforced on the types of equipment being used. The usual hours of working set by most Local Authorities are:

Monday to Friday	0800 to 1800
Saturday	0800 to 1300
Sunday and National Holidays	Nil

In the case of the erection of scaffolding for the works, then obviously these hours could not apply as one does not erect and strike scaffolding each and every day.

The 'usual' working hours for scaffolding would be twenty-four continuous and remaining.

'WHY' they may (not shall) enter is the part that is mostly misunderstood. It has been known for many surveyors to demand entry under section 8(2) in order that they might prepare a Schedule of Condition. This scenario arises because the Adjoining Owner refuses entry and in fact does not want anyone on his land at any time.

It seems quite powerful to be able to flex ones muscles, clutching a copy of the Act in your fist and requesting that a constable, by way of change, accompany you to the site in order to break an entry.

Confusion will naturally occur at the police station, but nevertheless after careful explanation and showing section 8(2) of the Act in black and white, you could have your constable assisting you with breaking and entering the Adjoining Owner's premises.

The Act gives us the reason why this may happen and says that it is *'for the purpose of executing any work in pursuance of the Act'*. Here is the hard bit, because the preparation of a Schedule of Condition is arguably not in pursuance of the Act. The Schedule of Condition is not even mentioned in the Act and it does not instruct that it should be undertaken.

It is however extremely prudent that a Schedule of Condition should be undertaken for the protection of both parties, and it is mostly done so, but not in the name of the Act, so be very careful with this one. An Adjoining Owner could sue and be successful if you break into his house just to prepare a Schedule of Condition.

Two
The Act says in section 8(1) that you may remove any furniture and fittings, or take any other action necessary for that purpose.

'A building owner, his servants, agents and workmen may during usual working hours enter and remain on any land or premises for the purpose of executing any work in pursuance of this Act and may remove any furniture or fittings or take any other action necessary for that purpose.'

Firstly it is assumed that the furniture and fittings are removed because they are in the way of the proposed works, such as the erection of scaffolding or protective screens. It is assumed that they may be moved out of the way, and we must assume carefully, without damage and protected in the new position.

The Act does not say, however that they should be put back afterwards but one could take it that common sense would apply and one should naturally do so.

Secondly that word *necessary* appears again, that is to say in the execution of work in pursuance of the Act, which as previously explained is an absolute and one should be extremely careful in applying this. If the work is therefore necessary and authorised with an award, then it must follow that the rights of entry must apply.

Three

The Act says in section 8(2) that you 'may' (not 'shall') *'break open any fences or doors in order to enter the premises'*. It does not say that you may enter through a window or any other means but by through fences or doors. So following the letter of the law, if the door is totally barricaded up but the window is ajar, you do not have the right to take the simple route of entry.

It is also interesting that the Act does not say what happens after you have forced an entry, but one assumes that you cannot just simply walk away and leave the premises laid open.

Entering through a door does not of course mean that a sledge hammer is to be used, attractive as may be for any adventurous boy on the right side of the law. The accompaniment of a locksmith will probably do the trick but not as much fun! The locksmith's bill would have to be paid by the Adjoining Owner if it is proven that he has thwarted the entry.

However if the Adjoining Owner is absent and not in the knowledge of his requirements, then any necessary bills and fees must fall to the Building Owner, as he is the one wishing to undertake the works.

Four

In the case of emergency the matter is much simpler and may be done as reasonably practicable: section 8(3)(a). This may be for the purpose of shutting off services, fire or hazard, or of course the saving of life.

Five

Any of the appointed surveyors may have the right of entry, but they must give fourteen days Notice, or less in an emergency. Usually this fourteen-day period is an impractical waiting time and invariably instant access is not usually denied, but be careful because it is not of right.

Conclusions

1. No absolute right of entry for Schedule of Condition.
2. Do only what is necessary.
3. Enter only through fence or door.
4. Emergency entry is permitted however gained, but reasonable notice required if possible.
5. Notice should be given to enter.

SECTION 9
Easements

A very short section, but an important one in that it states that the Act does not override easements.

> *'Nothing in this Act shall-*
>
> *(a) authorise any interference with an easement of light or other easements in or relating to a party wall; or*
>
> *(b) prejudicially affect any right of any person to preserve or restore any right or other thing in or connected with a party wall in case of the party wall being pulled down or rebuilt.'*

One

The subject of easements has been touched upon in section 1 and sometimes it is difficult to establish what is an easement and what is a trespass. Whilst this is not the duty of a party wall surveyor to determine such he clearly needs to highlight the potential problems and should be able to give advice on how to solve the matter.

To this end I believe that the party wall surveyor should have knowledge of other Acts and pieces of legislation that are connected to party wall matters, these are given as further reading in the Appendix.

More to the point, a knowledge of such will help in giving the best advice on how to overcome the problems that arise in undertaking party wall work, and therefore some guidance is given as follows:

Easement

An easement is a right benefiting a piece of land belonging to the dominant owner, to permit the use of that land in allowing certain and various actions over an adjoining piece of land belonging to the servient owner.

Such commonly known easements would be underground services, rights of way, rights of light and rights of support.

An easement can be created in many ways. One is by *'express grant'*, which is basically a Deed of Grant stating the terms of the easement. It can be found cited in a conveyance deed or a transfer deed.

Another way is by *'prescription'*, which happens when an act is undertaken repeatedly and continuously without the permission of the landowner for a period of twenty years. This time period needs to be sixty years when applying it to crown or church land.

An easement cannot be created by an unlawful act, for example walking across some ones land does not make it a right of way. However once an easement is established it is not easy to extinguish and generally would last forever. An easement can of course be changed by agreement between the parties. An example would be that an existing right of way across the back of a terrace of houses could be re-sited further away in order to allow building works to the rear of those houses. Notwithstanding that an easement *'runs with the land'* and is passed on with the land at any transfer stage.

Shared driveways are easements, which basically allows a right of way across each other person's land. These are commonly seen between pairs of semi-detached properties built in the 1930's. The boundary between the properties is usually down the middle of the common driveway between the properties. Use is made to pass over this shared driveway but not to build on it, nor indeed to park any vehicles on it, or place sheds or dustbins on it, insofar that this would block the use of the shared driveway for the other party.

Right of support is an easement and care should be taken with this one, as it quite often affects party walls. The demolition of one part of a pair of semi-detached houses could cause the collapse of its neighbour, remembering that not all party walls are of adequate thickness or of external quality building material. The right of support is applicable to retaining walls and excavation that is close to an adjoining property.

Right of light is another easement that arises in party wall work and is a complicated area. It should be understood however that the right is enjoyed by the land and not by a particular building containing a window. It is possible to demolish a building, build a replacement and still claim a right to light for the new windows, on the understanding of course that the new windows are not smaller than the previous ones.

The usual way that a right to light is obtained is by the Prescription Act 1832, which establishes the twenty-year rule. This twenty-year period must be continuous, and should it be broken for say one year, then the twenty-year period must start again from scratch - i.e. continuity is required.

Rights of light do not have to be registered on the title deeds of the property to become valid, although they can be so registered. Prevention of properties establishing a right to light can be gained by invoking the procedures set out under the Right to Light Act 1959.

There are mathematical ways of calculating the right to light, which are not for discussion here and not the subject of this book, but suffice to say the general established practice is that a room should receive light for 50% of its floor area at worktop level.

There are other claimed rights that have brought cases to court due to the blockage by a new building in front of an existing building, which have been lost. Such cases as the right to air for the use of a chimney or windmill, and the right to sunshine for solar power energy cases have also been lost. The only way to try and avoid such problems is to make objections at the planning stages, assuming that the proposals are covered by planning laws and not negated by permitted development rights. This may well change in the future as new energy sources are established and required.

Covenants

These are particular restrictions placed in the title of the property that need the agreement of another party before carrying out proposed works to the property. This is regardless of Town Planning, Building Regulations, Party Wall or any other Acts.

It is however possible to have covenants removed, and very often are removed by default. Section 84 of The Law of Property Act 1925 allows the removal of covenants by making an application to the Lands Tribunal. The Lands Tribunal Rules 1996 gives details of the requirements and information needed, who in turn will publish a Notice about the application.

Unlike Town Planning law, objections to the application can only be made by those actually affected and legally entitled to the benefit of the restriction. Compensation may be awarded to the objector for the loss of entitlement. If no objections are made then very often the Tribunal may determine the application without a hearing.

Conclusion

1. Be careful to establish if an easement or covenant exists, which could affect the work.
3. Check the charges register for easements and covenants etc.

SECTION 10
Resolution of dispute

This is the main part of the Act that deals with the Award. It is incumbent upon the surveyors to resolve disputes, and not to cause disputes. Disputes can only arise between owners as a result of Notices being served under sections 1, 2 or 6.

One

It is important to note that a dispute can only arise under the Act once a Notice is served. It cannot be emphasised enough that there is no dispute without a Notice, which is why many surveyors err in their official appointments, claiming for example that they have been appointed at the outset of the proposed works, and at the time they serve a Notice on behalf of the Building Owner.

As previously mentioned a surveyor cannot be appointed as a party wall surveyor unless a notice has been served and there is dissent, either expressly by the party on whom the notice is served, or the dissent is 'deemed' because of lack of response to the Notice.

It is very important to note that once a Notice has been served it does not mean that the whole procedure of section 10 must be carried through as of right or duty, simply because if the Owners agree on any matter at ANY time then the surveyors must abide by their decisions.

It has been known that surveyors have insisted on pursuing the matter to the serving of an Award regardless of the Owners wishes on the grounds that their appointment cannot be rescinded under section 10(2). Clearly this has to be utter nonsense and probably shows that such surveyors are merely fee driven and forgetting the reason for their statutory appointment, which is to resolves disputes between Owners. It is submitted that once a dispute between the Owners falls away, then so does the surveyors remit under the Act.

Two

When a Notice has been served, and dissent has arisen, which could be by actual notification or by lack of response, thus bringing about a deemed dissent, it is the first duty of both owners to appoint surveyors in order to resolve the dispute that has arisen between them.

What both owners can of course do, under section 10(1)(a), is to appoint the one surveyor between them who is known as the Agreed Surveyor. This often brings about an error that surveyors make with regards to the appointment of the agreed surveyor. This comes about when an Owner fails to appoint a surveyor and in accordance with the Act it is left for the other Owner to make an appointment on his behalf as given in section 10(4).

The Building Owner however is not at liberty to appoint one surveyor as the Agreed Surveyor, he must appoint another surveyor. The appointment of an Agreed Surveyor has to be made by the two owners, which is why he is so called – he is the surveyor that has been agreed between the parties. Section 10(1)(a) says that *'both parties shall concur in the appointment of one surveyor'*

This is borne out by section 10(6) because if a surveyor for example refuses to act effectively, the other surveyor may proceed as though he were the agreed surveyor. There are some surveyors in practice who get themselves appointed as the agreed surveyor this way but it is submitted that this is incorrect.

Surveyors must understand that they are not appointed to act for their respective appointing owners but that their appointment is one of statute, and that collectively, as a tribunal, they jointly make an

Award. This point was made clear in the case of *Frances Holland School v Wassef [2001] 2 EGLR 88*

The Act is quite clear in section 10(2) whereby all appointments shall be in writing and that such appointments cannot be rescinded.

Three

The Act is also clear as to the appointment of surveyors, and it goes on to say in section 10(1)(b) that the first thing that they must do is to select a third surveyor. Unless they do then a 'practical tribunal' would not have been formed, and as was the case in *Adams v Marylebone Borough Council [1907] 2 KB 822* the judge stated *'..........and the legislature has constituted a practical tribunal for deciding such questions.'* The judge in that case constantly referred to the 'statutory tribunal' throughout his summary. Without such a 'practical' tribunal in place (i.e. without a third surveyor selected) the appointed surveyors will have no remit.

This tribunal of surveyors can only determine matters that are related to the particular Notice served as was observed in the case of *Leadbetter v Marylebone Corporation [1904] 2 KB 893* and the same tribunal cannot determine any other matter, which occurred before their appointment was made. This was emphasised in *Woodhouse v Consolidated Property Corporation Ltd [199] 1 EGLR 174* and in *Louis v Sadiq*
[1997]1 EGLR 136

This is another place where one should take careful note insofar that it is the duty of both surveyors to 'select' and not 'appoint' a third surveyor. A common mistake seen in many awards is the clause that says *'in the event of his being unable or unwilling to act and the two appointed surveyors being unable to jointly agree upon a substitute, they have agreed that another third surveyor shall be appointed by the president of the Royal Institution of Chartered Surveyors'.*

The above 'standard clause' found in many Awards contains two errors, firstly that another surveyor must be 'selected' not 'appointed' and secondly the selection should be made in accordance with the Act, i.e. the Local Authority and not the president of any institution. Such a person is not mentioned in the Act.

Probably some of the confusion is caused because section 10(8) of the Act says that the selection is to be made by the *'appointing officer'*, who under section 20 is *'the person appointed under this Act by the Local Authority to make such appointments':* However, he is the person who selects the third surveyor under section 10(8)(a). Basically this is another example of bad drafting in the Act in using the word appoint instead of select.

The third surveyor only becomes appointed when he is requested by either party or either surveyor to determine a matter as given under section 10(11). His position at that stage changes from being selected to appointed, when therefore he becomes eligible for fees.

The appointment of the third surveyor also should be in writing and usually comes about by the fact that either of the parties or either of the surveyors approaches him formally.

The third surveyor does not have to give any reasons for his decision(s) but it is usually prudent to do so in order that the decision(s) so given will be clearly understood, and from whence comes further learning and knowledge.

My own advice is to *'select the third surveyor with care'* and not just someone well known who is quite often chosen. His position, as is the position of the other two surveyors, is by way of an 'expert determination', and is stated thus in the case of *Chartered Society of Physiotherapy v Simmonds Church Smiles [1995] 1 EGLR 155,* where the judge said '..... *an award under the Act is, in my judgement, sui generis and is more in the nature of an expert determination'*. 'sui generis' simply means 'the only member of a species', which is what many would say of party wall surveyors!

You may well wonder who the appointing officer is at the Local Authority. Well it may surprise you to know that many Local Authorities do not even know themselves. The general consensus is that it would be the Chief Building Inspector, and that is to whom I believe you should first go. The appointing officer could of course be someone else, maybe in the legal department.

Four
An appointed surveyor can only be removed from his position if he becomes 'incapable of acting', which in itself can only amount to a few reasons such as becoming ill, moving far away from the area, or some other legitimate reason.

An appointed surveyor cannot remove himself and deem himself 'incapable of acting' on the grounds that he does not get along with his appointing owner: section 10(5). However if he is the Agreed surveyor in section 10(3) he can refuse to act, and the reasons are not given. The whole procedure must therefore start again in order to settle the dispute.

Should the third surveyor be deemed incapable of acting the other two surveyors must select another surveyor forthwith: section 10(9)

There is of course nothing to stop a surveyor from getting out of his appointment by simply just refusing or neglecting to act, and therefore it would be left open for the other surveyor to act 'ex-parte'. It is considered, however, that if the surveyor neglects to act and his appointing Owner suffers as a consequence he could be sued for breach of statutory duty.

Five
There is a distinct difference between section 10(6) and 10(7) with regards to *refusing* to act and *neglecting* to act. If a surveyor *refuses* to act effectively the other surveyor may then act 'ex-parte' and no reason or notice needs to be given. If however a surveyor *neglects* to act for ten days then the other surveyor may act 'ex-parte'.

Therefore, should a surveyor have difficulty with the other surveyor he must be careful to ascertain the reason – is he refusing to act or neglecting to act? If he gets it wrong and proceeds 'ex-parte' his decision could invalidate the Award. It is therefore important to have written down, the reason for taking the 'ex-parte' action.

Both sections allow a surveyor to act *ex-parte* but note well only section 10(7) requires a ten-day Notice

Six
Clarification is required for section 10(10) as this is another misunderstood part. It states, *'The agreed surveyor or as the case may be the three surveyors or any two of them shall settle by award any matter-*

> *a) which is connected with any work to which this Act relates, and*
> *b) which is in dispute between the building owner and the adjoining owner.*

It is the reading of the words *'any matter'* and *'with any work to which this Act relates'*.

Many surveyors use it as a 'catch all' and is often used even for works that would have been notifiable but have since been completed before a Notice was served. As has been previously explained the Act is not invoked until a Notice is served.

For example a Building Owner commences work without a Notice and during this time damage is caused to the Adjoining Owner's property. The work stops and a Notice is served whereby the Adjoining Owner wishes to address the matter of compensation under the Act. He certainly cannot do so and the matter can only be resolved at common law.

If one takes the logic of retrospective actions and uses this sub-section for this purpose it would mean that even when the whole works are complete, and maybe some time has passed, a surveyor could take action under the Party Wall etc. Act 1996.

This is of course most certainly incorrect and a clear misunderstanding simply because the Act makes no provision for retrospective works. Indeed it could not do so; otherwise service of a Notice would be superfluous. The Act is written around a Notice.

Seven

It is the duty of the party wall surveyors to inform their appointing owners that the owners may call upon the third surveyor to determine a disputed matter as written in section 10(11). This should be done as soon as the third surveyor has been selected.

It may well be of course that surveyors do not wish their decisions challenged and taken to the third surveyor, but the Act is quite equitable in this respect, and surveyors should not withhold knowledge of such rights.

Eight
The first paragraph of section 10(12) says, *'An award may determine-*

....

c) any other matter arising out of or incidental to the dispute including the costs of making the award.'

It should be noted that 'may' is used and not 'shall'. An Award would be quite legitimate if no mention was made of fees, but I do not know of any surveyor who would leave such out such a clause, and there is very good reason for the fees to be included in the Award as I will explain.

Everything in an Award becomes legally binding within fourteen days of it being served, assuming of course that it is not proven to be *ultra vires*. This is a commonly used expression and simply means *'beyond the powers'*, which is an act that falls outside or beyond the jurisdiction of a court or individual authority.

An Award can be enforced in court if any breach has occurred. Therefore this means that any fee due to a surveyor for making the Award is far more enforceable if it is contained in the Award. This is why I would advocate the inclusion of both surveyors' fees and not just the Adjoining Owner's surveyor's fees as is usually done.

I never understand why the Building Owner's surveyor's fee is not normally stated in the Award, is it because they do not want the Adjoining Owner or his surveyor to know the level of fee that he is charging? I have to say by way of response that very often the Adjoining Owner's surveyor's fee has been known to be higher! (probably because they know that the Building Owner will have to pay)

The Building Owner's surveyor has undoubtedly quoted his fee to the Building Owner at the outset, probably before he was appointed.

Problems always seem to arise around the matter of fees, and in my dealings with some surveyors; it seems to be the only thing on their minds. Of course everyone wants to get paid for the work they do but I would stress here that this is not the purpose of the Act. Surveyors should remember that they are appointed to administer and implement the Act in order to resolve a dispute between the Owners.

It is considered to be illegal to hold an Award in lien against fees as once an Award has been agreed and signed by the surveyor(s) the Award must be served forthwith on the Appointed Owners as stated in section 10(14).

There is no reason why a surveyor should not be paid immediately upon the service of an Award, but he should not be paid before it is released. Many awards erroneously state that *'immediately upon signing of the award the building owner shall pay £xxx'*. Unless the Building Owner is there at the time of the signing of the award, and it is highly unlikely that he would be, he cannot comply with this clause. The wording should say *'immediately upon serving'*, not signing.

Section 10(13) of the Act deals with matter of reasonable costs in the preparation of an Award, and, apart from noting that they should be reasonable a fee scale cannot be given here.

A Building Owner can at the outset, agree fees with a surveyor before appointing him, but he has no real control over the fee charged by the surveyor appointed by the Adjoining Owner. Suffice to say that it must be reasonable and that the fee stated in an Award payable to such a surveyor could be appealed. Many would see this to be a little unfair on the Building Owner, but he is the one that who is wishing to do the work so it is only right that he should pay all the fees.

The Act does of course say in section 10(13) that it is the surveyor or surveyors who will determine who should pay the reasonable costs incurred in dealing with the matter. It is of course more usual for the surveyor(s) to agree that the costs be borne by the Building Owner. Indeed it has almost become an unwritten rule, but it is not actually specified as such in the Act.

It clearly has to be this way, since the Building Owner could thwart the choice of the Adjoining Owner's appointment of a surveyor until he is satisfied with the fee he will have to pay. As the Building Owner wishes to carry out the works he must serve a Notice and there is then the right of the Adjoining Owner to appoint a surveyor of his choice. The Building Owner can have no say in this part, and rightly so, as it would be like questioning the appointment of another person's solicitor when attempting to resolve a dispute between parties.

The level of fee claimed by the surveyor to the Adjoining Owner is usually the problem in that it can be considered to be too high. The key word in the Act is *'reasonable'*, a word used in law, but what is reasonable? The short answer is common sense. That is to say, how long will it really take to undertake this work in man-hours that can be justified?

By way of example let us assume that two surveyors meet on site for one hour, usually to prepare a schedule of condition. Then let us assume that a draft Award is sent to the other surveyor for comment, how long will it honestly take (an experienced surveyor) to read a few pages and comment – half an hour? Bear in mind that the Award usually contains many standard acceptable clauses.

Now let us assume one further site inspection is to be allowed during the works and one final inspection for completion to check off the schedule of condition, although my experience is that most surveyors do not undertake the latter. Nevertheless, it has been allowed for so let us add another two hours at most. Then add the paperwork and telephone calls for about another hour. So then in truth we only have about 4-5 hours.

Too simplistic you may say, let us then add another two or three hours and maybe call it a day's work at eight hours, then multiply this by your hourly charge out rate and you have your *'reasonable fee'*.

It is very wise to keep a good paper trail on all projects and party wall work is no exception to this. Keep a written note of the time expended on the job and if it is questioned, proof can be given.

The above is of course a simple case, but the principle is there to see, so any suspicion of a high fee can be questioned and refused if needs be. The fee element should on no account hold up an Award, but often does. An Award could be served *ex-parte* without the inclusion of a fee element on the basis that the other surveyor has refused to act.

A trip to the third surveyor brings problems on this issue because he will require his fee first and he is unlikely to determine wholly on one side, unless of course the question is so phrased to him that he must agree or disagree with the fee.

When surveyors are appointed well outside of the geographical area of the work, and the Owners are entitled to do this, it is nevertheless unreasonable to expect to pay additional costs to cover travelling etc. It is also unreasonable to have to pay urban rates for provincial works, particularly if the works are of a minor nature.

This is notwithstanding that some works are complex and need the input of an experienced surveyor who may well be from outside of the area, and therefore clearly acceptable under such conditions. However, it is considered advisable to have appointed surveyors from the geographical area of the work, or at least reasonably close to it whenever possible.

Nine

As soon as the Award has been made and agreed it must be served on the appointing owners: section 10(14), and the moment it is served the Building Owner would be at liberty to commence the works if he so chooses.

A very common mistake made by surveyors is caused by the misunderstanding of section 10(17) where the mention of fourteen days is made, in which either of the owners may appeal the Award in the county court.

The common clause that is used in Awards reads as follows:

> *'That fourteen days after the delivery of the signed Award the Building Owner shall be at liberty if he so chooses....................'*

Such a clause is saying that the work must not commence until fourteen days has passed. There is no provision in the Act to write such a clause. The mention of fourteen days is the period of time allowed for the owners to appeal the Award in the county court. It does not mean that the Building Owner or his contractor has to wait fourteen days. Certainly he could be at risk if the Award is challenged within those fourteen days, but in the vast majority of cases the Award is never taken to court under an appeal.

It is most unreasonable to make the Building Owner and his contractor wait fourteen days before work can start, in fact I believe it is unlawful to do so, and strangely enough the Award could be appealed on this clause alone and the Building Owner would succeed.

The ramifications of holding up the start of the works for fourteen days on a large expensive project, whereby the contactor is under contract could be put into jeopardy. Notwithstanding that it could have taken some time to prepare the Award.

Ten

The conclusiveness of the Award is final, as stated in section 10(17) even to the extent of it not being able to be questioned in any court. This is taken to be after the fourteen-day period of section 10(17) has expired. However an Award can be challenged after that date if it is considered to be *ultra vires*. The case of *Stone and Hastie [1903] 2 KB 463* and *Gyle Thompson v Wall Street (Properties) Ltd [1974] 1 All ER 295.* establishes the fact that invalid Awards do not necessarily need to be appealed in the court.

As previously mentioned an Award is a legal document and therefore it has to be prepared with care, for if it is not it could be challenged in the court. An Award may be deemed *ultra vires* if it tries to deal with matters that are outside of the Act, or is made without jurisdiction. An Award can be challenged if it is wrong 'on the face of it', for example the recitals give the names of the owners or surveyors wrongly.

It must be remembered that surveyors draw up an Award as an agreement between two owners because a dispute has arisen between the owners. On this basis it is only possible for an Award to be challenged by the owners and never the surveyors as some might think.

The challenge is made in the county court by the appellant (one owner appealing the Award) and responded to by the respondent (other owner). This is the right given to them under section 10(17) of the Act. The legal costs are a matter for the owners and not the surveyors. Such an appeal is a 'statutory' appeal and should therefore be made under Part 52 of the Civil Procedure Rules (CPR). This follows from the decision of the court in *Zissis v Lukomski & Anor [2006] EWCA Civ 341*

It should be noted however, that if Part 52 of the CPR applies to appeals under section 10(17) of the Act, this throws into doubt the fourteen-day time period allowed by section 10(17) since the Practice Direction for Part 52 at paragraph 17.3 allows twenty-eight days for the filing of statutory appeals. It has been considered however that the fourteen days stated in the Act would take precedence over the CPR rule of twenty-eight days.

If however an owner believes that the surveyor(s) have been negligent and/or failed in their statutory duty, a separate action could be taken against the surveyor(s). So while the party wall surveyor has great powers under the Act and to a greater extent is protected by it, common sense will tell you that the matter is still open to law. The party wall surveyor is not so protected like arbitrators or adjudicators.

The obvious advice is to be careful how the Award is drawn up and ensure that the contents are correct and legally valid, otherwise surveyors will be costing the owners money, who will rightly seek redress. It is worth mentioning again here the words of Mr Justice Brightman in the well known Gyle Thompson case '*.........It therefore seems to me important that the steps laid down by the Act should be scrupulously followed throughout, and short cuts are not desirable..............the approach of surveyors to those requirements ought not to be casual'*.

Eleven
A very common occurrence is when a Building Owner starts the work without serving a Notice and the Adjoining Owner does not know what to do and makes contact with a party wall surveyor. The surveyor begins to act for the Adjoining Owner and uses the Act to sort out the trouble. Unfortunately there is nothing that can be done by using the Act, because as previously mentioned the Act only becomes invoked when a Notice is served.

The only real advice to an Adjoining Owner when an erring Building Owner will not listen is to first put into writing his complaint and thereafter apply for an injunction to stop the works. The Adjoining Owner and not the surveyor must do this.

An injunction as defined in section 25 of the CPR is *'a court order prohibiting a person from doing something or requiring a person to do something'*. This is meant to be an equitable remedy to resolve a matter of conflict. The party that fails to adhere to the injunction could face civil or contempt of court proceedings, and may have to pay damages, or suffer sanctions for such failure.

Injunctions are a way of relief that cannot be resolved by monetary damages, and can be temporary or permanent. It is usual in most cases to have an interim injunction until such time that the court makes a final decision. The interim injunction is therefore seen as a preventive measure to guard against further damage, rather than a remedy for past events. Injunctions have long been used against nuisances.

In section 36(6) of the Building Act 1984, an express provision is made for the right of *'any other person'* to make an application to the High Court for an injunction to be made if the works contravene the Building Regulations. So if the works also contravene the Building Act 1984 then this would add weight to formulating reasons for an injunction to be given.

Needless to say, the prospect may on the face of it seem quite daunting but in reality getting an injunction can be quite easy providing it is applied for properly. One has to simply give good reasons to a judge to convince him to serve an injunction.

Some questions to consider in making an application could be:

1. What is the reason for an injunction? – It will not be given simply because a Notice under the Act has not been served.
2. What will be the legal costs involved as it could well be that they will not be all claimed back?
3. What damage has been caused or is likely to be caused?
4. What loss will be incurred?
5. How will you be affected if there were no injunction?
6. Is an injunction a 'proportionate' step for the court to take?

As to the costs of applying for an injunction, these need not be much at all if the application is made by the Adjoining Owner aided with some salient points, or better still a report, formulated by a surveyor. A copy of a standard *'Application for Injunction'* is in the appendix.

Assuming that an injunction is given, the judge could give such directions that the Building Owner should serve Notice under the Act, at which stage the surveyors can begin to do their duty in accordance with the Act to set the matter right.

There is of course the possibility that a surveyor could strongly guide a Building Owner into serving a Notice informing him that it might well be the cheapest option. The surveyor however, if successful, can only bring in the Act at that point and cannot retrospectively cover any of the work that has already been undertaken.

A Notice cannot be served after the event to bring about the workings of the Act. Any such Award that might be made covering previous works would be deemed to be invalid. If the works have in fact been completed, the best that could be done is to try and agree some remedy for the damages, but this would be outside of the Act and merely dealt with at common law.

The logic is quite clear in that the reason for serving a Notice is to inform the Adjoining Owner of intended works covered by the Act that are to be carried out. It therefore follows that if the work has already been done then any form of retrospective Notice is useless. Despite this, some surveyors contrive to serve a retrospective Notice followed by a retrospective Award for which the Act makes no provision. *Gyle-Thompson and Others v Wall Street (Properties) Ltd [1974] 1 All ER 295*

Conclusions

1. Only a Notice can invoke section 10.
2. Surveyors' appointments must be in writing
3. A third surveyor must be selected by the other two surveyors.
4. You cannot release yourself from your appointment easily.
5. Understand the difference between refusing and neglecting to act.
6. No retrospective works can be made in an Award.
7. Surveyors ought to inform the Owners about the role of the third surveyor.
8. Only agree a fee that is reasonable.
9. An Award must be served as soon as it is signed.
10. Only Owners can challenge an Award by way of appeal.
11. Consider carefully the matter of injunctions.

SECTION 11
Expenses

One

If damage occurs to an Adjoining Owner's property during the works and is found to be attributable to those works, the Adjoining Owner would obviously wish it to be repaired, and Section 11 (8) deals with this aspect.

Section 11(8) of the Act allows for monies to be paid in lieu of carrying out the repair works, and very often this course is taken, as quite often confidence is lost by the Adjoining Owner with the Building Owner's contractor by the mere fact that he has caused such damage.

Very often the two surveyors will agree a sum of money by way of settlement, but the best way is to obtain actual estimates from builders who would undertake the work, as this will be the likely real cost of putting everything right. The Adjoining Owner can take the money but is under no obligation to repair the damage.

Two

The sub-section that seems to cause most concern is section 11(11) whereby payment is due for the use of an existing party wall.

'Where use is subsequently made by the adjoining owner of work carried out solely at the expense of the building owner the adjoining owner shall pay a due proportion of the expenses incurred by the building owner in carrying out that work; and for this purpose he shall be taken to have incurred expenses calculated by reference to what the cost of the work would be if it were carried out at the time when that subsequent use is made.'

Basically the Act is saying that if use is made of a party wall that was built by a Building Owner then the Adjoining Owner (who becomes a Building Owner) shall make a payment for using the wall.

The problem that some surveyors and owners have is who pays whom, and it is not a question as to who actually built the wall in the first place. The Act is clear when it says *'Where use is subsequently made by the adjoining owner of work carried out solely at the expense of the building owner the adjoining owner shall pay a due proportion of the expenses incurred by the building owner in carrying out that work...........'*

The *'subsequent use'* may be many years later, but the right passes with the freehold. It has in fact been tested at law in the case of *Mason v Fulham Corporation [1910] 1 KB 631*. However in the case of *Stone and Hastie, Re[1903] 2 KB 463* the money due was not paid to the tenant of the property who had a 21 year term, but to the freeholder who had originally raised the wall.

Nevertheless it is established that payment must be made for the use of a party wall and the cost of the works are to be at today's prices and not the cost of when the wall was first built. The Act states *' and for this purpose he shall be taken to have incurred expenses calculated by reference to what the cost of the work would be if it were carried out at the time when the subsequent use was made'*.

What should be included in the compilation of the cost of the wall is often an issue between the owners, and their surveyors. I believe the answer is everything that would be required to build the wall, were it a project.

This therefore would mean the cost of labour, materials, preliminaries, professional fees and VAT etc. This might also include the cost of underpinning had it been necessarily undertaken at the time of a previous use of the wall by a previous owner.

Most surveyors would cost the works and divide equally to achieve the payable sum, but care should be taken with this as the Act simply says *'a due proportion'* so it does not have to be 50%, although in practice it usually is.

There is of course a situation whereby one owner could make a payment to another at the time of construction. This would have to be legally drafted and appended to the deeds of the respective parties, as proof would need to be shown at a later date.

It is important to understand what can happen upon change of ownership, as this affects the validity of a Notice, and works that have commenced before and after change of ownership. If the adjoining ownership changes prior to work commencing then a new Notice would need to be served upon the new owner, as they may wish to consent or dissent accordingly. If the works have already begun then the new owner is bound by the Notice and Award, and has no opportunity to contest unless within the fourteen day period.

The reader should also acquaint himself with the cases of *Carlash v Salt [1906] 1 Ch 335, Selby v Whitbread & Co. [1917] 1 KB 736* and *Observatory Hill Ltd v Camtel Investments SA [1997] 1 EGLR 140* with regards to change of ownership.

However one could say by way of compensation that the new owner would be allowed the benefit of the payment from the Building Owner for the use of the wall if that portion of the works had not yet been reached. The general rule is that benefits can be assigned but burdens cannot not be so assigned.

Three

Section 11(11) deals with the future use of a party wall and it is probably the best place to deal with the law of trespass. That is in the case where an owner makes use of a boundary wall and presumes to make it a party wall.

This is a situation where time is a primary factor insofar that if the use can be proven to have been made for more than twenty years then prescriptive rights (Prescription Act 1832) would allow the situation to remain.

If the use can be proven to have been made for more than ten years then adverse possession could be claimed and an application made to Land Registry. The owner of the boundary wall would of course have the right to challenge such a claim and unlike pre-Land Registry Act 2000, adverse possession would not be automatically granted.

If the use was less than ten years the owner of the wall could require any attachment to be removed, and have the wall returned to its former condition, as the use would be simple trespass. However if the owner of the wall is happy with the use and is not disposed to deprive his neighbour of his previous enjoyment, he can agree to the use but only on the grounds of a revocable licence.

The warning to owners then is if you do not want to lose part of your property, ensure that the neighbour's use is only temporary and under licence. Such an agreement should be appended to the property deeds.

Conclusions

1. Settle damages with estimates
2. Include all that would be necessary to build the wall.
3. Determine what is trespass and what is permitted.

SECTION 12
Security for expenses

One

Section 12 tends to be used only for larger contracts, and is not generally used in relation to domestic works.

The works may be quite extensive and costly, therefore the Adjoining Owner needs to have assurances that the works will be completed and that he is not left 'holding the baby' so to speak in that he will have to complete the works in order to protect his own property.

By virtue of section 12(1) the Adjoining Owner could request some funds to be set-aside in a separate account to be used in the event that the Building Owner, due to liquidation, insolvency or negligence does not complete the works.

Such monies would be agreed between the surveyors in the form of an Escrow Account but only if the parties are unable to agree themselves. A model of an Escrow account appears in the appendix.

An Escrow is a deed held by a third party taking effect only when a stated condition is fulfilled. It is a deposit or fund held in a trust as a security. One would not normally require such security if the Building Owner is a large public body such as a Local Authority.

There could of course be other ways to give security such as a bond, personal guarantees, or a one off project insurance cover. I believe the best way is by actually having funds in an account that can be quickly accessed if needs be, which is why an Escrow account is the better format for security.

The surveyors cannot determine the amount of the security, that is for the owners to agree, but of course should they not agree then the surveyors could help, as this would be considered a dispute between the owners. It is also not the duty of the surveyors to make judgements about the financial position of the owners.

This security could of course be requested for any size of project if there may be doubt about the Building Owners ability to complete the works, for example the Building Owner may be undertaking the works on a DIY basis or is employing seemingly 'cowboy builders'.

The converse is also applicable by virtue of section 12(2) against the Adjoining Owner if he has served a counter notice for works to be undertaken and there is a doubt as to his probability of settling the account. The Building Owner could therefore ask for security for expenses.

The security for expenses can of course only relate to that works which would be notifiable and in pursuance of the Act.

If there is a disagreement about the Escrow sum then it follows that an Award cannot be served, simply because the Escrow must be in place before an Award is served. The reason for this is that under section 10(14) when an Award is signed it must be served forthwith on the parties. This means that as soon as the Building Owner receives the Award he would be at liberty to commence the works, thus nullifying the Escrow.

An Award that is served without an Escrow in place is likely to be challenged in court.

Conclusion

1. Escrow figures are to be agreed between owners where possible.

SECTION 13
Account for work carried out.

This section is linked with section 11, which deals with the recovery of expenses to which the Adjoining Owner is liable to meet.

Any work that the Building Owner has done that benefits the Adjoining Owner for which payment is due, then the Building Owner must render an account.

The Act lays down time limits for this and says that the account must be rendered within two months of the completion of the works. The account must be in writing and give such details particularising the work, which should be valued and set at fair average rates.

When the account has been served on the Adjoining Owner he has just one month in which to raise any objections by way of a Notice, otherwise the account shall have been deemed to be accepted.

Should the Adjoining Owner raise an objection to the account then the Act says that a dispute is deemed to have arisen. It would therefore seem to follow that such a dispute is resolved under section 10 by the appointed surveyor(s).

It is considered that these time scales are important to follow and should not be taken lightly. It is therefore incumbent upon the surveyor(s) to inform both owners of these time scales.

Conclusion.

1. Accounts for expenses should be met within the time scales.

SECTION 14
Settlement of account.

This short section deals with the resolution of payment. The Act says that the ownership of the works carried out rests with the Building Owner until the account is settled by the Adjoining Owner.

Clearly it can be seen that any use that the Adjoining Owner makes of the works before payment would probably be considered a trespass. Similarly it is considered that unless the account is rendered within the time frames laid down in section 13 then the Building Owner could lose his right to possession of title.

Conclusion

1. The Adjoining Owner should be made aware, that use cannot be made of the works carried out by the Building Owner, until payment has been made.

SECTION 15
Service of notices etc.

Section 15 is the most important part of the Act because without a Notice the Act cannot be invoked. It is therefore very important that Notices should be properly served. The content of the Notice has been discussed under section 3 and this section deals with the actual service of the same.

One
The formal requirement for serving of a Notice or other document such as an Award can principally be in four ways:

1. Delivering in person to the Adjoining Owner.
2. Posting to the usual address of the Adjoining Owner, home or business or last known residence.
3. Delivering or posting to the secretary or clerk in the case of a corporate body at its registered or principal office.
4. Affixing the Notice to the premises of the owner if no one can be found at that address.

The name of the recipient does not have to be given, and merely addressed to 'the owner', although it is usually best to try and find such a name. A Notice would not be invalid if it simply says 'the owner', but would be if the Building Owner's name and address were not given. All Building Owners must be named, however many, whereas just one Adjoining Owner can be named in order for a Notice to be valid.

Care should be taken with personal delivery of a Notice or an Award as handing it to an agent or another party would be seen as not served. This is akin to the service of writs that must be personally handed over. In the case of serving a Notice or an Award on a public body, the safest form therefore would be by post.

Conclusion

1. Ensure the Notice and Award is properly served.

SECTION 16
Offences

This section reinforces the rights of access in section 8, which basically says that failure to comply means that he or she is guilty of an offence.

The guilty person is liable on summary conviction in the magistrates' court to a fine of an amount not exceeding level three on the standard scale. This level three is currently set at £1000.

SECTION 17
Recovery of sums

This is just a two line section but enough to let you know that the surveyors fees are easily recoverable because of this section. As previously mentioned it is always wise to include the fees of both surveyors in the Award, that way it will be more enforceable than simply having to chase for fees in the normal way.

At this point I should draw to your attention that the word *'fees'* does not occur in the Act but are always referred to as *'costs'*.

The usual way that surveyors take to recover their fees is in the county court. The court will give direction for payment without question, simply because the Act effectively says so in section 10(16) in that the Award shall not *'be questioned in any court.'*

When this section says 'summarily' that means via the magistrates' court. Therefore this section provides the additional right to pursue a debt in the magistrates' court, over and above the usual county court.

This unusual way, that is actually more effective and quicker, is through the magistrates' court. This court may be approached because the surveyor has a statutory appointment. It is also more effective because much of the general public has become complacent with the county court system and knows that even after being awarded the judgement, the claimant may still have trouble in collecting his money.

The magistrates' court is likely to provide a much quicker hearing and its power to bring defendants to court for non-compliance is much greater with penalties. This is because the magistrates' court is primarily known as a criminal court, although it does have power to deal with civil debts, such as with the enforcement of council tax, gas and electric supplies, social security and tax arrears.

An application to a magistrates' court is by way of 'complaint', which is governed by section 8 of the Magistrates' Court Act 1980. The magistrate can award costs, unlike the county court for 'small claims', and failure to meet the order for payment can be enforced by the issue of a distress warrant. Contrary to some views, imprisonment is not within the magistrates' power for default in this particular instance.

The claimed sum needs to be more than £50,000 to be brought in the High Court, which of course is generally unlikely with regards to surveyor's fees.

Surveyors fees are civil debts because they are owed to an agent of the State. The party wall surveyor has a statutory appointment and as such falls within this category.

Problems always seem to arise around the matter of fees, and in personal dealings with some surveyors it would seem to be the only thing on their minds. Of course everyone wants to get paid for the work they do, but I would stress here that this is not the purpose of the Act. Surveyors should remember that they are appointed to administer and implement the Act in order to resolve a dispute between the owners.

Conclusion

1. You are entitled to pursue any sum payable 'summarily' in the magistrates' court.

SECTION 20
Interpretation

The meaning of 'surveyor' has been questioned by some as who might be *'party to the matter'*. Some have seen this as being the architect or engineer or designer etc.

This is on the basis that such a person could not be truly neutral or impartial in his dealings. For example if the engineering aspect is questioned and the other appointed surveyor wanted to call for another opinion to check the details, the designer engineer acting as party wall surveyor is going to defend his scheme quite vehemently.

It would of course be quite natural for him to do so, otherwise what does he say when his fee for designing has been met by his client but yet further fees are incurred by the designer because they are questioned by the other surveyor?

It such a circumstance the third surveyor could be called upon to decide and rule in the matter, which is rather unfortunate because this would probably mean unnecessary increased costs to the owner.

Nevertheless the Act is quite clear and has in fact been clarified at law in the case of *Loost v Kremer [1997] (unreported) 12 May* when the judge said *'........the fact that he is an architect does not seem to me to invalidate his appointment. There is no evidence of conflict of interest,.............it must frequently happen that parties in this situation do appoint persons who have acted for them because they know very much more about the building than any new appointee would..........The mere fact that he has acted as architect does not, in my judgement mean that he must disqualify himself................he changes his capacity from being simply an agent to a quasi arbitrator and he has to bear in mind that those are his duties..................in no way does his previous capacity prevent him from acting in the new capacity'.*

One could however say that such a heavily involved designer should make the wise decision and not undertake the position of being the party wall surveyor, but clearly the Act allows him to be appointed.

What is of course true is that an experienced party wall surveyor cannot act for himself if he becomes either the Building Owner or the Adjoining Owner, for he would be *'party to the matter'*. Any such surveyor would be champing at the bit!

Finally

All experience is good, as hopefully lessons are learnt from mistakes of what people call bad experiences. Without mistakes one does not learn, as we are commonly told. Some indignant surveyors claim that *'I have been doing it this way for 25 years'* to which one can only reply that they have been doing it wrongly for 25 years.

The author has to confess that he was one such surveyor who like others was relying upon his predecessors with more experience to be correct in all that they did and wrote. Then one day he read the Act for himself and to his disgrace discovered he had been making mistakes for many years – and somehow had gotten away with it!

If surveyors enjoy party wall work then they will want to get it right as much as they can, bearing in mind there will always be some differences of opinion for certain 'grey' areas, and this must be respected. It is recommended that surveyors do their homework, building up their knowledge and experiences.

There is a caveat however and that is to remember that we are surveyors and not lawyers, have a good knowledge of the laws pertaining to and around the Act but leave the finer arguments to those qualified to do so.

The closing words therefore are: *'Let the Act be your guide, add nothing and take nothing away from the contents and you will not go wrong'*. An obvious statement to make perhaps, but if surveyors heed it well they will not end up in court, at least not on the wrong side!

APPENDIX

A further reading list:

The Arbitration Act 1996
Boundaries - Halsbury's Laws of England.
Access to Neighbouring Land Act 1992
Anti-Social Behaviour Act 2003 - part 8 High Hedges
Landlord & Tenant (Covenants) Act 1995
Prescription Act 1832
Limitation Act 1980
Law of Property Act 1925 and 1969
Land Registration Act 2002
Rights of Light & Waldram Test
Town and Country Planning Act 1990
Building Regulations 2006
Gale on Easements:
    Rights of Light
    Rights of Way
    Rights of Support
Commonhold and Leasehold Reform Act 2002
Control of Pollution Act 1974 Section 60-62 – [Noise]
Environmental Protection Act 1990
Health and Safety at Work Act 1974
Construction & Design Management (Amendment) Regulation 2000 – [Site Safety]

SCAFFOLD LICENCE

Name of Building Owner..

Address..

....

Name of Adjoining Owner..

Address..

....

The Building Owner wishes to undertake building works to his property that adjoins the Adjoining Owner's property and in order to undertake such works requires the temporary use of the Adjoining Owner's land to support scaffold.

The necessity for such is for ease of construction in providing a safe working platform and to comply with the Health and Safety Regulations.

IT IS HEREBY AGREED

1.00 **That** the sole costs of the erection and maintenance of such scaffolding shall be borne by the Building Owner.

2.00 **That** the plan numbered.................shall show the dimensional extent of the required scaffold.

3.00 **That** the scaffold shall remain in place for the duration of the necessary works to be undertaken commencing (date)........................... and shall be dismantled immediately at completion of the works (date)..............................or for an extended period only by written agreement.

4.00 **That** the scaffold shall be erected by a trained licensed scaffolder and be so maintained during its stay of erection.

5.00 **That** the scaffold shall be fitted with tarpaulins and screens so as to protect the Adjoining Owner's property from dust, debris and for privacy.

6.00 **That** the scaffold shall be fitted with an appropriate alarm system to provide protection for Adjoining Owner's property against intruders.

7.00 **That** the scaffold shall be fitted with a lighting system, hazard tapes and warning signs for visual protection.

8.00 **That** a payment shall be made in advance by the Building Owner to the Adjoining Owner for the use of the said land in the allowance for the erection of the said scaffold in the sum of £.............. which covers the period specified above. An additional payment may be due if this period is extended.

9.00 **That** in the event of a dispute due to damage or any other associated cause the parties shall refer the matter to (Name of Surveyor).......................................

10.00 **That** the times of working on the scaffold shall be limited to 0800 to 1800 hours Monday to Friday and 0800 to 1300 hours on Saturday. No work shall be permitted using the scaffold on Sundays or Public/National holidays.

11.00 **That** the Building Owner or his contractor is to hold Public Liability insurance policy for the sum of not less than

£1,000,000. A copy of such certificate to be produced and given to the Adjoining Owner.

12.00 **That** at the end of the works when the scaffold is removed the Building Owner or his contractor shall clean down any surfaces, gutter and gullies belonging to the Adjoining Owner so affected by the works, and to remove any fallen debris, leaving the Adjoining Owner's land and property clean and tidy.

13.00 **That** the contractor shall not be permitted access to any parts beyond the agreed scaffolding area.

14.00 **That** any breach of this agreement shall allow the Adjoining Owner to have the scaffolding removed forthwith, with the full costs of any reinstatement to be borne by the Building Owner.

15.00 **That** the Building Owner shall pay any legal or surveyors fees due in the preparation of this licence.

Building Owner....................................

Adjoining Owner................................. Date..................

Witness...

Address...

....

CRANE LICENCE

Name of Building Owner..

Address..

....

Name of Adjoining Owner..

Address..

....

The Building Owner wishes to undertake building works to his property that adjoins the Adjoining Owner's property and in order to undertake such works requires the temporary use of the Adjoining Owner's air space to over-sail with a crane.

IT IS HEREBY AGREED

1.00 **That** the Building Owner will erect a (type of crane).......................on his own land in order to facilitate the construction of his building works.

2.00 **That** the plan numbered.................shall show the dimensional position of the required crane, the extent of its boom, and its over-sail dimension.

3.00 **That** the crane shall remain in place for the duration of the necessary works to be undertaken commencing (date)........................... and shall be dismantled immediately at completion of the works (date)...............................or for an extended period only by written agreement.

4.00 **That** the crane shall be erected by a trained licensed crane erector and be so maintained during its stay of erection.

5.00 **That** no load carried by the crane shall be suspended over the Adjoining Owner's property.

6.00 **That** the lowest part of the crane over-sailing the air space shall be not less than 5 Metres above the Adjoining Owner's air space.

7.00 **That** if any interference should occur to electrical or telephone facilities due to the use of the crane then the Building Owner shall make provision for such remedy.

8.00 **That** a payment shall be made in advance by the Building Owner to the Adjoining Owner for the use of the said air space in the allowance for the erection of the said crane in the sum of £.............. which covers the period specified above. An additional payment may be due if this period is extended.

9.00 **That** in the event of a dispute due to damage or any other associated cause the parties shall refer the matter to (Name of Surveyor).......................................

10.00 **That** the times of working on the crane shall be limited to 0800 to 1800 hours Monday to Friday and 0800 to 1300 hours on Saturday. No work shall be permitted using the crane on Sundays or Public/National holidays.

11.00 **That** the Building Owner or his contractor is to hold Public Liability insurance policy for the sum of not less than

£1,000,000. A copy of such certificate to be produced and given to the Adjoining Owner.

12.00 **That** at the end of the works when the crane is removed the Building Owner or his contractor shall check that nothing has been affected at ground level or where the crane has over-sailed the Adjoining Owner's building, and rectify accordingly.

13.00 **That** any breach of this agreement shall allow the Adjoining Owner to have the crane removed forthwith, with the full costs of any reinstatement to be borne by the Building Owner.

14.00 **That** the Building Owner shall pay any legal or surveyors fees due in the preparation of this licence.

Building Owner....................................

Adjoining Owner.................................. Date..................

Witness..

Address...

....

ESCROW AGREEMENT

Name of Building Owner...

Address...

....

Name of Adjoining Owner...

Address...

....

The Building Owner wishes to undertake building works to his property that adjoins the Adjoining Owner's property and in order to undertake such works the Adjoining Owner requires that a sum of money be deposited as security against the works.

IT IS HEREBY AGREED

1.00 **That** the two appointed surveyors under the Party Wall etc. Act 1996 have agreed to act as experts in this matter.

2.00 **That** the Building Owner will deposit the sum of £.......with (name and address of agent).

....

....

....

who is/are to act as the Escrow Agent.

3.00 **That** the sum of money shall be deposited before the works commence and in accordance with clause ....of the Award that has been served on both parties.

4.00 **That** if a dispute arises that would cause the invocation of the Escrow Monies then this is to be settled by the two surveyors, or in the event that they cannot agree, it shall be put before the third surveyor for a determination.

5.00 **That** if a dispute does arise then the Adjoining Owner and his agent will be at liberty to enter the Building Owner's premises in order to undertake the necessary works.

6.00 **That** the monies held by the Escrow Agent will be released for the purposes directed by the surveyor(s) if so required.

7.00 **That** if the monies held by the Escrow Agent are found to be insufficient to undertake the completion of the necessary works then the works will be completed by the Adjoining Owner who will be at liberty to recover the monies from the Building Owner.

8.00 **That** the monies held by the Escrow Agent will be held in an interest bearing account and repaid to the Building Owner in the event that the monies are not used.

9.00 **That** the Escrow Agents fees shall be paid by the Building Owner.

10.00 **That** the Escrow Agreement shall come to an end upon completion of the works, as advised by the surveyor(s) and any or all-remaining monies shall be returned to the Building Owner and the account closed.

Signed

Building Owner...Date.....................

Adjoining Owner...Date....................

Witness...

Address...

....
.

FORM 98

COMPLAINT

(MC Act 1980,ss51, 52; MC Rules 1981, r 4)

[insert name of court] Magistrates' Court

Date:	*[insert date]*
Defendant:	*[insert debtor's name]*
Address:	*[insert debtor's address]*
Matter of Complaint:	Sum recoverable pursuant to section 17 Party Wall etc Act 1996 ("the Act")

<u>Particulars</u>

- The Complainant is the Building Owners' / Adjoining Owners' / Agreed / Third *[delete as appropriate]* Surveyor and was so appointed by virtue of s.10 (1)(a / b*[delete as appropriate]*) of the Act.
- In pursuance of the said appointment, the Complainant has undertaken works as follows:

[Insert brief details of work conducted]

- By Award dated *[insert date]* the Complainant's fees of £*[insert amount]* were payable by the Defendant, in accordance with s.10 (12)(c) of the Act. *[delete this paragraph if payment of fees is not covered by Award]*
- The Complainant has served invoices upon the Defendant for his fees in the sum of £*[insert amount]*. To date, the Defendant has not paid the Complainant's fees.

The Complainant therefore seeks recovery of the said sum of £*[insert amount]* as a civil debt, pursuant to section 17 of the Act, plus interest on the said sum at 8%, plus the costs of proceedings, currently £*[insert amount for costs of proceedings to date]*

The Complaint of: *[insert your name]*
Address: *[insert your address]*
Telephone: *[insert your telephone number]*

Who (upon oath) states that the defendant was responsible for the matter of complaint of which particulars are given above.

Taken (and sworn) before me,

(Justice of the Peace)
..(Justices' Clerk)

Application for Injunction (General Form)

In the	
County Court	
Claim No.	
Claimant's Ref.	
Defendant's Ref.	

Between ________________________ ☐ Claimant ☐ Applicant ☐ Petitioner *(Tick whichever applies)*

and ________________________ ☐ Defendant ☐ Respondent

Notes on completion

Tick whichever box applies and specify legislation where appropriate

(1) Enter the full name of the person making the application

(2) Enter the full name of the person the injunction is to be directed to

(3) Set out here the proposed restraining orders (if the defendant is a limited company delete the wording in brackets and insert "whether by its servants, agents, officers or otherwise")

(4) Set out here any proposed mandatory orders requiring acts to be done

(5) Set out here any further terms asked for including provision for costs

(6) Enter the names of all persons who have sworn affidavitsor signed statement in support of this application

(7) Enter the names and addresses of all persons upon whom it is intended to serve this application

(8) Enter the full name and address for service and delete as required

☐ By application in pending proceedings

☐ Under Statutory provision ________________________

Seal

The Claimant (Applicant/Petitioner)(1)

applies to the court for an injunction order in the following terms:

That the Defendant (Respondent)(2)

be forbidden (whether by himself or by instructing or encouraging any other person(3)

And that the Defendant(Respondent)(4)

And that(5)

The grounds of this application are set out in the written evidence of(6) sworn (signed) on

This written evidence is served with this application.

This application is to be served upon(7)

This application is filed by(8)

(the Solicitors for) the Claimant (Applicant/Petitioner)

whose address for service is

Signed Dated

This section to be completed by the court

* Name and address of the person application is directed to

To*

of

This application will be heard by the (District) Judge

at

on the day of [19][20] at o'clock

If you do not attend at the time shown the court may make an injunction order in your absence

If you do not fully understand this application you should go to a Solicitor, Legal Advice Centre or a Citizens' Advice Bureau

STANDARD MODEL AWARD

This is a standard model award for use when two surveyors are appointed.

An award under the provisions of Party Wall etc. Act 1996 to be served forthwith on the appointing owners under section 10 (14)

Whereas, *(name of building owner)* whose address is *(address of building owner)*
(Hereinafter referred to as the **Building Owner**), owner within the meaning of the said **Act** of the premises *(address of premises)*
Did on the *(date of notice)* serve upon
(name of adjoining owner) of *(address of adjoining owner)*
(Hereinafter known as the **Adjoining Owner**), owner within the meaning of the **Act** of the adjoining premises.
Notice of *his/her/their* intention to exercise the rights given to *him/her/them* by the **Party Wall etc. Act 1996** Section *(state relevant sections of the Act)* by executing the works more particularly defined in the notice.

And Whereas a dispute has arisen between the building owner and the adjoining owner (known as the parties) the building owner has appointed *(name and address of surveyor)* to act as *his/her/their* surveyor and the adjoining owner has appointed *(name and address of surveyor)* to act as *his/her/their* surveyor.
And Whereas the two surveyors so appointed have selected *(name*

and address of surveyor) to act as third surveyor in accordance with the provisions of the **Act** and, in the event of *his/her* being unable or unwilling to act and the two surveyors being unable to jointly agree upon a substitute, they have agreed that another third surveyor shall be selected in accordance with section 10(8) of the Act.

Now we, being the two surveyors so appointed, having inspected the said premises, **Do Hereby Award and Determine** as follows: -

1.

(a) **That** the wall *(describe the wall - party, boundary etc)* is sufficient for the purposes of the adjoining owner.

(b) **That** the drawings numbered *(insert drawing numbers).* attached hereto and signed by the two surveyors forms part of this award.

(c) **That** a schedule of condition *(dated)* attached hereto and signed by the two surveyors forms part of this award.

(d) **That** photographs have been taken of the adjoining owners property and will be held on file for record purposes.

2. **That** after the service of the signed award the building owner shall be at liberty if *he/she/they* so choose*(s),* carry out the following works: -

(a) *(describe the works)*

3. **That** no material deviation from the agreed works shall be made without prior consultation with and agreement by the two surveyors.

4. **That** if the building owner exercises the above rights *he/she/they* shall:-

   (a) Execute the whole of the aforesaid works at the sole cost of the building owner.

   (b) Take all reasonable precautions and provide all necessary support to retain the adjoining owner's land and buildings.

   (c) Make good forthwith all structural or decorative damage to the adjoining owner's property occasioned by the said works in materials to match existing to the satisfaction of the two surveyors, or if so required by the adjoining owner make payment in lieu of carrying out the works to make good the damage.

(d) Hold the adjoining owner free from liability in respect of any injury or loss of life to any person or damage to property caused by, or in consequence of, the execution of the said works, and the costs of making any justified claims.

(e) Permit the two surveyors to have access to the building owner's premises and the adjoining owner's premises at all reasonable times during the progress of the said works.

(f) Carry out the whole of the said works so far as is practical from the building owner's side.

(g) Ensure that adequate insurance is held for the works and to cover for any possible claim(s) that may arise from any party.

(h) *(add any other necessary clauses)*

5. **That** the whole of the works referred to in this award shall be executed in accordance with the Building Regulations, and any other requirements of statutory authorities, and shall be executed in a proper and workmanlike manner in sound and suitable materials in accordance with the terms of this award to the reasonable satisfaction of the two surveyors.

6. **That** the works shall be carried through with reasonable expedition after commencement and so as to avoid unnecessary inconvenience to the adjoining owners or occupiers. In particular, noisy works shall be restricted to between the hours of 0800 and 1800 on Weekdays and 0800 and 1300 on Saturdays. No works shall be permitted on Sundays or National Holidays. *(amend times as necessary)*

7. **That** the signed awards shall be served forthwith on the appointing owners. A copy of the award will be retained by the two surveyors. A copy of the signed award shall be

provided to the building owner's contractor with instructions to incorporate the provisions within his work.

8. **That** the building owner shall pay at the service of this award the adjoining owner's surveyor's fee of £.......... in connection with the preparation of this award and for *(number of intermediate inspection of the works),* and for a final inspection of the works. In the event of damage being caused or other contingencies or variations arising, a further fee shall be payable at the rate of £.............. per hour.

9. **That** the building owner shall pay at the service of this award the building owner's surveyor's fee of £.......... in connection with the preparation of this award and *(number of intermediate inspection of the works),* and for the final inspection of the works. In the event of damage being caused or other contingencies or variations arising, a further fee shall be payable at the rate of £.............. per hour.

10. **That** the said surveyors reserves the right to make and serve any further award or awards that may be necessary, as provided in the said **Act**.

11. **That** nothing in this award shall be held as conferring, admitting or affecting any right to light or air, or any other easements or covenants.

12. **That** any agreement or acceptance made by the surveyors in this award or subsequently during the works on site shall not be taken to imply any responsibility by them or their appointed technical delegates for structural or any other insufficiency in any part of the works whether existing or executed.

IN WITNESS WHEREOF having set their hands this ....day of (month) (year)

Surveyor to the building owner...

Witness

Name...

Address...

....

Occupation..

Surveyor to the adjoining owner..

Witness

Name..

Address...

....

Occupation..

NEW MODEL AWARD

It is considered that the standard awards that have been used for many years need to be updated and should be written in clear English for all to understand especially for the appointing owners, which after all is the whole purpose of preparing such awards.

An award should therefore not be continued to be written in such English and style that is out of date and not in keeping with modern English usage of the language. There should not be anything mysterious about an award, or that it should give the impression that it needs to be written by 'quasi- lawyers'.

There is the well known Plain English Campaign launched in 1979 and has a Crystal Mark that was produced in 1990 which is awarded to organisations that produce documents in simple plain English.

Here then is a model award when used by two surveyors that has been set out in clear English.

An Award under the provisions of the Party Wall etc Act 1996 which is to be served on the Building Owner and the Adjoining Owner.

Name of Building Owner...

Address...

....

Name of Adjoining Owner..

Address...

....

Site Address...

....

Date of Notice..

Sections of the Act covered by the Notice..

Name and address of surveyor appointed by the Building Owner.

....

....

Name and address of surveyor appointed by the Adjoining Owner.

....

....

Name and address of surveyor selected to be the Third Surveyor.

....

....

(If the selected surveyor is unable or unwilling to act then another surveyor shall be selected in accordance with section 10(8) of the Act).

1.00 The two surveyors have inspected the premises and have agreed the following;

1.01 The wall separating the building owner and the adjoining owner is a party wall as defined in the Act.

1.02 A schedule of condition has been taken with photographs and records the present condition of the wall and premises. The schedule is dated......................and is appended to this award.

1.03 The drawings numbered............................have been prepared by ....and are appended to this award.

2.00 After this award has been served the building owner may carry out the following works.

2.01 (List the works)..

....

....

3.00 Any changes to the agreed works shall be notified to the two surveyors in writing.

4.00 The works shall be at the sole cost and risk of the building owner unless otherwise agreed.

5.00 The building owner shall take all reasonable precautions to protect the adjoining owner's property, and to provide adequate supports to ensure the safety and stability of the adjoining owner's property during the course of the works.

6.00 The building owner shall remove any scaffolding and screens that have been erected for the works as soon as the work applicable to the notice has been completed.

7.00 The building owner shall make good any damage so caused to the adjoining owner's property to the satisfaction of the two surveyors, or if so required by the adjoining owner that a payment shall be made in lieu of carrying out the works, which shall be agreed by the two surveyors.

8.00 The adjoining owner's property shall be kept free of debris and cleaned at the end of every working day. On completion of the works the building owner shall ensure that all gutters and gullies are cleaned, and all surfaces so affected by the works shall be cleaned.

9.00 The building owner shall hold the adjoining owner free from any liability in respect of any injury or loss of life to any third party claims or damage to property caused by the works.

10.00 The building owner, his workmen and party wall surveyor shall be permitted access to the adjoining owner's premises as necessary at all reasonable times during the progress of the works. Reasonable notice to be given as required.

11.00 The whole of the works as covered by the notice shall be executed in accordance with the Building Regulations and any other statutory requirements.

12.00 The works shall proceed with diligence to completion so as to avoid unnecessary inconvenience to the adjoining owner. Noisy work shall be restricted to the hours of 0800 to 1800 on weekdays and 0800 to 1300 on Saturdays. No work shall be permitted on Sundays or National Holidays.

13.00 The signed awards shall be served forthwith on the building owner and adjoining owner. A copy of the award shall be held in the files of the two surveyors and a further copy shall be given to the building owner that he may inform his contractor of the relevant content.

14.00 The building owner shall pay on service of the award the following costs to the two surveyors.

Building owner's surveyor..............£.........plus VAT
and at an hourly rate of £..........plus VAT should further involvement be necessary.

Adjoining owner's surveyor............£........plus VAT
and at an hourly rate of £................plus VAT should further involvement be necessary.

15.00 The two surveyors reserve the right to serve further awards that may be necessary.

16.00 The two surveyors do not accept any responsibility for any structural or other insufficiency of the works.

17.00 This award is made without prejudice to the rights of any other person or bodies having an interest in the party wall.

Building Owner Surveyor..Date....................

Witness...

Address...

....

Adjoining Owner Surveyor.......................................Date................

Witness..

Address...

....

Party Wall etc. Act 1996

1996 Chapter 40

ARRANGEMENT OF SECTIONS

Construction and repair of walls on line of junction

Party Wall etc. Act 1996

1996 CHAPTER 40

An Act to make provision in respect of party walls, and excavation and construction in proximity to certain buildings or structures; and for connected purposes. [18th July 1996]

BE IT ENACTED by the Queen's most Excellent Majesty, by and with the advice and consent of the Lords Spiritual and Temporal, and Commons, in this present Parliament assembled, and by the authority of the same, as follows:-

Construction and repair of walls on line of junction

New building on line of junction.
1.-(1) This section shall have effect where lands of different owners adjoin and-

(a) are not built on at the line of junction; or
(b) are built on at the line of junction only to the extent of a boundary wall (not being a party fence wall or the external wall of a building),

and either owner is about to build on any part of the line of junction.

(2) If a building owner desires to build a party wall or party fence wall on the line of junction he shall, at least one month before he intends the building work to start, serve on any adjoining owner a notice which indicates his desire to build and describes the intended wall.

(3) If, having been served with notice described in subsection (2), an adjoining owner serves on the building owner a notice indicating his consent to the building of a party wall or party fence wall-

(a) the wall shall be built half on the land of each of the two owners or in such other position as may be agreed between the two owners; and
(b) the expense of building the wall shall be from time to time defrayed by the two owners in such proportion as has regard to the use made or to be made of the wall by each of them and to the cost of labour and materials prevailing at the time when that use is made by each owner respectively.

(4) If, having been served with notice described in subsection (2), an adjoining owner does not consent under this subsection to the building of a party wall or party fence wall, the building owner may only build the wall-

(a) at his own expense; and
(b) as an external wall or a fence wall, as the case may be, placed wholly on his own land,

and consent under this subsection is consent by a notice served within the period of fourteen days beginning with the day on which the notice described in subsection (2) is served.

(5) If the building owner desires to build on the line of junction a wall placed wholly on his own land he shall, at least one month before he intends the building work to start, serve on any adjoining owner a notice which indicates his desire to build and describes the intended wall.

(6) Where the building Owner builds a wall wholly on his own land in accordance with subsection (4) or (5) he shall have the right, at any time in the period which-

(a) begins one month after the day on which the notice mentioned in the subsection concerned was served, and
(b) ends twelve months after that day,

to place below the level of the land of the adjoining owner such projecting footings and foundations as are necessary for the construction of the wall.

(7) Where the building owner builds a wall wholly on his own land in accordance with subsection (4) or (5) he shall do so at his own expense and shall compensate any adjoining owner and any adjoining occupier for any damage to his property occasioned by-

(a) the building of the wall;
(b) the placing of any footings or foundations placed in accordance with subsection (6).

(8) Where any dispute arises under this section between the building owner and any adjoining owner or occupier it is to be determined in accordance with section 10.

Repair etc. of party wall: rights of owner.
2.-(1) This section applies where lands of different owners adjoin and at the line of junction the said lands are built on or a boundary wall, being a party fence wall or the external wall of a building, has been erected.

(2) A building owner shall have the following rights-

(a) to underpin, thicken or raise a party structure, a party fence wall, or an external wall which belongs to the building owner and is built against a party structure or party fence wall;

(b) to make good, repair, or demolish and rebuild, a party structure or party fence wall in a case where such work is necessary on account of defect or want of repair of the structure or wall;

(c) to demolish a partition which separates building belonging to different owners but does not conform with statutory requirements and to build instead a party wall which does so conform;

(d) in the case of buildings connected by arches or structures over public ways or over passages belonging to other persons, to demolish the whole or part of such buildings, arches or structures which do not conform with statutory requirements and to rebuild them so that they do so conform;

(e) to demolish a party structure which is of insufficient strength or height for the purposes of any intended building of the building owner and to rebuild it of sufficient strength or height for the said purposes (including rebuilding to a lesser height or thickness where the rebuilt structure is of sufficient strength and height for the purpose of any adjoining owner);

(f) to cut into a party structure for any purpose (which may be or include the purpose of inserting a damp course);

(g) to cut away from a party wall, party fence wall, external wall or boundary wall any footing or any projecting chimney breast, jamb or flue, or other projection on or over the land of the building owner in order to erect, raise or underpin any such wall or for any other purpose;

(h) to cut away or demolish parts of any wall or building of an adjoining owner overhanging the land of the building owner or overhanging a party wall, to the extent that it is necessary to cut away or demolish the parts to enable a vertical wall to be erected or raised against the wall or building of the adjoining owner;

(j) to cut into the wall of an adjoining owner's building in order to insert a flashing or other weather-proofing of a wall erected against that wall;

(k) to execute any other necessary works incidental to the connection of a party structure with the premises adjoining it;

(l) to raise a party fence wall, or to raise such a wall for use as a party wall, and to demolish a party fence wall and rebuild it as a party fence wall or as a party wall;

(m) subject to the provisions of section 11(7), to reduce, or to demolish and rebuild, a party wall or party fence wall to –

(i) a height of not less than two metres where the wall is not used by an adjoining owner to any greater extent than a boundary wall; or

(ii) a height currently enclosed upon by the building of an adjoining owner;

(n) to expose a party wall or party structure hitherto enclosed subject to providing adequate weathering.

(3) Where work mentioned in paragraph (a) of subsection (2) is not necessary on account of defect or want of repair of the structure or wall concerned, the right falling within that paragraph is exercisable-

(a) subject to making good all damage occasioned by the work to the adjoining premises or to their internal furnishings and decorations; and

(b) where the work is to a party structure or external wall, subject to carrying any relevant flues and chimney stacks up to such a height and in such materials as may be agreed between the building owner and the adjoining owner concerned or, in the event of dispute, determined in accordance with section 10;

and relevant flues and chimney stacks are those which belong to an adjoining owner and either form part of or rest on or against the party structure or external wall.

(4) The right falling within subsection (2)(e) is exercisable subject to-

(a) making good all damage occasioned by the work to the adjoining premises or to their internal furnishings and decorations; and

(b) carrying any relevant flues and chimney stacks up to such a height and in such materials as may be agreed between the building owner and the adjoining owner concerned or, in the event of dispute, determined in accordance with section 10;

and relevant flues and chimney stacks are those which belong to an adjoining owner and either form part of or rest on or against the party structure.

(5) Any right falling within subsection (2)(f),(g) or (h) is exercisable subject to making good all damage occasioned by the work to the adjoining premises or to their internal furnishings and decorations.

(6) The right falling within subsection (2)(j) is exercisable subject to making good all damage occasioned by the work to the wall of the adjoining owner's building.

(7) The right falling within subsection (2)(m) is exercisable subject to-

(a) reconstructing any parapet or replacing an existing parapet with another one; or

(b) constructing a parapet where one is needed but did not exist before.

(8) For the purposes of this section a building or structure which was erected before the day on which this Act was passed shall be deemed to conform with statutory requirements if it conforms with the statutes regulating buildings or structures on the date on which it was erected.

Party structure notices.

3.- (1) Before exercising any right conferred on him by section 2 a building owner shall serve on any adjoining owner a notice (in this Act referred to as a "party structure notice") stating-

(a) the name and address of the building owner;

(b) the nature and particulars of the proposed work including, in cases where the building owner proposes to construct special foundations, plans, sections and details of construction of the special foundations together with reasonable particulars of the loads to be carried thereby; and

(c) the date on which the proposed work will begin.

(2) A party structure notice shall-

(a) be served at least two months before the date on which the proposed work will begin;

(b) cease to have effect if the work to which it relates-

(i) has not begun within the period of twelve months beginning with the day on which the notice is served; and
(ii) is not prosecuted with due diligence.

(3) Nothing in this section shall-

(a) prevent a building owner from exercising with the consent in writing of the adjoining owners and of adjoining occupiers any right conferred on him by section 2; or

(b) require a building owner to serve any party structure notice before complying with any notice served under any statutory provisions relating to dangerous or neglected structures.

Counter notices.

4.-(1) An adjoining owner may having been served with a party structure notice serve on the building owner a notice (in this Act referred to as a "counter notice") setting out-

(a) in respect of a party fence wall or party structure, a requirement that the building owner build in or on the wall or structure to which the notice relates such chimney copings, breasts, jambs or flues, or such piers or recesses or other like works, as may reasonably be required for the convenience of the adjoining owner;

(b) in respect of special foundations to which the adjoining owner consents under section 7(4) below, a requirement that the special foundations-

(i) be placed at a specified greater depth than that proposed by the building owner; or

(ii) be constructed of sufficient strength to bear the load to be carried by columns of any intended building of the adjoining owner,

or both.

(2) A counter notice shall-

(a) specify the works required by the notice to be executed and shall be accompanied by plans, sections and particulars of such works; and

(b) be served within the period of one month beginning with the day on which the party structure notice is served.

(3) A building owner on whom a counter notice has been served shall comply with the requirements of the counter notice unless the execution of the works required by the counter notice would-

(a) be injurious to him;

(b) cause unnecessary inconvenience to him; or

(d) cause unnecessary delay in the execution of the works pursuant to the party structure notice.

Disputes arising under sections 3 and 4.

5. - If an owner on whom a party structure notice or a counter notice has been served does not serve a notice indicating his consent to it within the period of fourteen days beginning with the day on which the party structure notice or counter notice was served, he shall be deemed to have dissented from the counter notice and a dispute shall be deemed to have arisen between the parties.

Adjacent excavation and construction.

6.- (1) This section applies where-

(a) a building owner proposes to excavate, or excavate for and erect a building or structure, within a distance of three metres measured horizontally from any part of a building or structure of an adjoining owner; and

(c) any part of the proposed excavation, building or structure will within those three metres extend to a lower level than the level of the bottom of the foundations of the building or structure of the adjoining owner.

(2) This section also applies where-

(a) a building owner proposes to excavate, or excavate for and erect a building or structure, within a distance of six metres measured horizontally from any part of a building or structure of an adjoining owner; and

(b) any part of the proposed excavation, building or structure will within those six metres meet a plane drawn downwards in the direction of the excavation, building or structure of the building owner at an angle of forty-five degrees to the horizontal from the line formed by the intersection of the plane of the level of the bottom of the foundations of the building or structure of the adjoining owner with the plane of the external face of the external wall of the building or structure of the adjoining owner.

(3) The building owner may, and if required by the adjoining owner shall, at his own expense underpin or otherwise strengthen or safeguard the foundations of the building or structure of the adjoining owner so far as may be necessary.

(4) Where the buildings or structures of different owners are within the respective distances mentioned in subsection (1) and (2) the owners of those buildings or structures shall be deemed to be adjoining owners for the purposes of this section.

(5) In any case where this section applies the building owner shall, at least one month before beginning to excavate, or excavate for and erect a building or structure, serve on the adjoining owner a notice indicating his proposal and stating whether he proposes to underpin or otherwise strengthen or safeguard the foundations of the building or structure of the adjoining owner.

(6) The notice referred to in subsection (5) shall be accompanied by plans and sections showing-

(a) the site and depth of any excavation the building owner proposes to make;

(b) if he proposes to erect a building or structure, its site.

(7) If an owner on whom a notice referred to in subsection (5) has been served does not serve a notice indicating his consent to it within the period of fourteen days beginning with the day on which the notice referred to in subsection (5) was served, he shall be deemed to have dissented from the notice and a dispute shall be deemed to have arisen between the parties.

(8) The notice referred to in subsection (5) shall cease to have effect if the work to which the notice relates-

(a) has not begun within the period of twelve months beginning with the day on which the notice was served; and

(b) is not prosecuted with due diligence.

(9) On completion of any work executed in pursuance of this section the building owner shall if so requested by the adjoining owner supply him with particulars including plans and sections of the work.

(10) Nothing in this section shall relieve the building owner from any liability to which he would otherwise be subject for injury to any adjoining owner or any adjoining occupier by reason of work executed by him.

Rights etc. Compensation etc.

7.- (1) A building owner shall not exercise any right conferred on him by this Act in such a manner or at such time as to cause unnecessary inconvenience to any adjoining owner or to any adjoining occupier.

(2) The building owner shall compensate any adjoining owner and any adjoining occupier for any loss or damage which may result to any of them by reason of any work executed in pursuance of this Act.

(3) Where a building owner in exercising any right conferred on him by this Act lays open any part of the adjoining land or building he shall at his own expense make and maintain so long as may be necessary a proper hoarding, shoring or fans or temporary construction for the protection of the adjoining land or building and the security of any adjoining occupier.

(4) Nothing in this Act shall authorise the building owner to place special foundations on land of an adjoining owner without his previous consent in writing.

(5) Any works executed in pursuance of this Act shall-

(a) comply with the provisions of statutory requirements; and

(b) be executed in accordance with such plans, sections and particulars as may be agreed between the owners or in the event of dispute determined in accordance with section10;

and no deviation shall be made from those plans, sections and particulars except such as may be agreed between the owners (or surveyors acting on their behalf) or in the event of dispute determined in accordance with section 10.

Rights of entry

8.- (1) A building owner, his servants, agents and workmen may during working hours enter and remain on any land or premises for the purpose of executing any work in pursuance of this Act and may remove any furniture or fittings or take any other action necessary for that purpose.

(2) If the premises are closed, the building owner, his agent and workmen may, if accompanied by a constable or other police officer, break open any fences or doors in order to enter the premises.

(3) No land or premises may be entered by any person under subsection (1) unless the building owner serves on the owner and the occupier of the land or premises-

(a) in case of emergency, such notice of the intention to enter as may be reasonably practicable;

(b) in any other case, such notice of the intention to enter as complies with subsection (4).

(4) Notice complies with this subsection if it is served in a period of not less than fourteen days ending with the day of the proposed entry.

(5) A surveyor appointed or selected under section 10 may during usual working hours enter and remain on any land or premises for the purpose of carrying out the object for which he is appointed or selected.

(6) No land or premises may be entered by a surveyor under subsection (5) unless the building owner who is a party to the dispute concerned serves on the owner and the occupier of the land or premises-

(a) in case of emergency, such notice of the intention to enter as be reasonably practicable;

(b) in any other case, such notice of the intention to enter as complies with subsection (4).

Easements.
9.- Nothing in this Act shall-

(a) authorise any interference with an easement of light or other easements in or relating to a party wall; or

(b) prejudicially affect any right of any person to preserve or restore any right or other thing in or connected with a party wall in case of the party wall being pulled down or rebuilt.

Resolution of disputes
10.- (1) Where a dispute arises or is deemed to have arisen between a building owner and an adjoining owner in respect of any matter connected with any work to which this Act relates either-

(a) both parties shall concur in the appointment of one surveyor (in this section referred to as an "agreed surveyor"); or

(b) each party shall appoint a surveyor and the two surveyors so appointed shall forthwith select a third surveyor (all of whom are in this section referred to as "the three surveyors").

(2) All appointments and selections made under this section shall be in writing and shall not be rescinded by either party.

(3) If an agreed surveyor-

(a) refuses to act;

(b) neglects to act for a period of ten days beginning with the day on which either party serves a request on him;

(c) dies before the dispute is settled; or

(d) becomes or deems himself incapable of acting,

the proceedings for settling such dispute shall begin *de novo.*

(4) If either party to the dispute-

(a) refuses to appoint a surveyor under subsection (1)(b), or

(b) neglects to appoint a surveyor under subsection (1)(b) for a period of ten days beginning with the day on which the other party serves a request on him,

the other party may make the appointment on his behalf.

(5) If, before the dispute is settled, a surveyor appointed under paragraph (b) of subsection (1) by a party to the dispute dies, or becomes or deems himself incapable of acting, the other party who appointed him may appoint another surveyor in his place with the same power and authority.

(6) If a surveyor-

(a) appointed under paragraph (b) of subsection (1) by a party to the dispute; or

(b) appointed under subsection (4) or (5),

refuses to act effectively, the surveyor of the other party may proceed to act *ex parte* and anything so done by him shall be as effectual as if he had been an agreed surveyor.

(7) If a surveyor-

(a) appointed under paragraph (b) of subsection (1) by a party to the dispute; or

(b) appointed under subsection (4) or (5),

neglects to act effectively for a period of ten days beginning with the day on which either party or the surveyor of the other party serves a request on him, the surveyor of the other party may proceed to act *ex parte* in respect of the subject matter of the request and anything so done by him shall be as effectual as if he had been an agreed surveyor.

(8) If either surveyor appointed under subsection (1)(b) by a party to the dispute refuses to select a third surveyor under subsection (1) or (9), or neglects to do so for a period of ten days beginning with the day on which the other surveyor serves a request on him-

(a) the appointing officer; or

(b) in cases where the relevant appointing officer or his employer is a party to the dispute, the Secretary of State,

may on the application of either surveyor select a third surveyor who shall have the same power and authority as if he had been selected under subsection (1) or subsection (9).

(9) If a third surveyor selected under subsection (1)(b)-

(a) refuses to act;

(b) neglects to act for a period of ten days beginning with the day on which either party or the surveyor appointed by either party serves a request on him; or

(c) dies, or becomes or deems himself incapable of acting, before the dispute is settled,

the other two of the three surveyors shall forthwith select another surveyor in his place with the same power and authority.

(10) The agreed surveyor or as the case may be the three surveyors or any two of them shall settle by award any matter-

(a) which is connected with any work to which this Act relates, and

(b) which is in dispute between the building owner and the adjoining owner.

(11) Either of the parties or either of the surveyors appointed by the parties may call upon the third surveyor selected in pursuance of this section to determine the disputed matters and he shall make the necessary award.

(12) An award may determine-

(a) the right to execute any work;

(b) the time and manner of executing any work; and

(c) any other matter arising out of or incidental to the dispute including the costs of making the award;

But any period appointed by the award for executing any work shall not unless otherwise agreed between the building owner and the adjoining owner begin to run until after the expiration of the period prescribed by this Act for service of the notice in respect of which the dispute arises or is deemed to have arisen.

(13) The reasonable costs incurred in-

(a) making or obtaining an award under this section;

(b) reasonable inspections of work to which the award relates; and

(c) any other matter arising out of the dispute,

shall be paid by such of the parties as the surveyor or surveyors making the award determine.

(14) Where the surveyors appointed by the parties make an award the surveyors shall serve it forthwith on the parties.

(15) Where an award is made by the third surveyor-

(a) he shall, after payment of the costs of the award, serve it forthwith on the parties or their appointed surveyors; and

(b) if it is served on their appointed surveyors, they shall serve it forthwith on the parties.

(16) The award shall be conclusive and shall not except as provided by this section be questioned in any court.

(17) Either of the parties to the dispute may, within the period of fourteen days beginning with the day on which an award made under this section is served on him, appeal to the county court against the award and the county court may-

(a) rescind the award or modify it in such manner as the court thinks fit: and

(b) make such order as to costs as the court thinks fit.

Expenses

11.- (1) Except as provided under this section expenses of work under this Act shall be defrayed by the building owner.

(2) Any dispute as to responsibility for expenses shall be settled as provided in section 10.

(3) An expense mentioned in section 1(3)(b) shall be defrayed as there mentioned.

(4) Where work is carried out in exercise of the right mentioned in section 2(2)(a), and the work is necessary on account of defect or want of repair of the structure or wall concerned, the expenses shall be defrayed by the building owner and the adjoining owner in such proportion as has regard –

(a) the use which the owners respectively make or may make of the structure or wall concerned; and

(b) responsibility for the defect or want of repair concerned, if more than one owner makes use of the structure or wall concerned.

(5) Where work is carried out in exercise of the right mentioned in section 2(2)(b) the expenses shall be defrayed by the building owner and the adjoining owner in such proportion as has regard to –

(a) the use which the owners respectively make or may make of the structure or wall concerned; and

(b) responsibility for the defect or want of repair concerned, if more than one owner makes use of the structure or wall concerned.

(6) Where the adjoining premises are laid open in exercise of the right mentioned in section 2(2)(e) a fair allowance in respect of disturbance and inconvenience shall be paid by the building owner to the adjoining owner or occupier.

(7) Where a building owner proposes to reduce the height of a party wall or party fence wall under section 2(2)(m) the adjoining owner may serve a counter notice under section 4 requiring the building owner to maintain the existing height of the wall, and in such case the adjoining owner shall pay to the building owner a due proportion of the cost of the wall so far as it exceeds-

(a) two metres in height; or

(b) the height currently enclosed upon by the building of the adjoining owner.

(8) Where the building owner is required to make good damage under this Act the adjoining owner has a aright to require that the expenses of such making good be determined in accordance with section 10 and paid to him in lieu of the carrying out of work to make the damage good.

(9) Where –

(a) works are carried out, and

(b) some of the works are carried out at the request of the adjoining owner or in pursuance of a requirement made by him,

he shall defray the expenses of carrying out the works requested or required by him.

(10) Where –

(a) consent in writing has been given to the construction of special foundations on land of an adjoining owner; and

(b) the adjoining owner erects any building or structure and its cost is found to be increased by reason of the existence of the said foundations,

the owner of the building to which the said foundations belong shall, on receiving an account with any necessary invoices and other supporting documents within the period of two months beginning with the day of the completion of the work by the adjoining owner, repay to the adjoining owner so much of the cost as is due to the existence of the said foundations.

(11) Where use is subsequently made by the adjoining owner of work carried out solely at the expense of the building owner the adjoining owner shall pay a due proportion of the expenses incurred by the building owner in carrying out that work; and for this purpose he shall be taken to have incurred expenses calculated by reference to what the cost of the work would be if it were carried out at the time when that subsequent use is made.

Security for expenses

12.- (1) An adjoining owner may serve a notice requiring the building owner before he begins any work in the exercise of the rights conferred by this Act to give such security as may be agreed between the owners or in the event of dispute determined in accordance with section 10.

(2) Where –

(a) in the exercise of the rights conferred by this Act an adjoining owner requires the building owner to carry out any work the expenses of which are to be defrayed in whole or in part by the adjoining owner; or

(b) an adjoining owner serves a notice on the building owner under subsection (1).

The building owner may before beginning the work to which the requirement or notice relates serve a notice on the adjoining owner requiring him to give such security as may be agreed between the owners or in the event of dispute determined in accordance with section 10.

(3) If within the period of one month beginning with –

(a) the day on which a notice is served under subsection (2); or

(b) in the event of dispute, the date of the determination by the surveyor or surveyors,

the adjoining owner does not comply with the notice or the determination, the requirement or notice by him to which the building owner's notice under that subsection relates shall cease to have effect.

Account for work carried out.

13.- (1) Within the period of two months beginning with the day of the completion of any work executed by a building owner of which the expenses are to be wholly or partially defrayed by an adjoining owner in accordance with section 11 the building owner shall serve on the adjoining owner an account in writing showing-

(a) particulars and expenses of the work; and

(b) any deductions to which the adjoining owner or any other person is entitled in respect of old materials or otherwise;

and in preparing the account the work shall be estimated and valued at fair average rates and prices according to the nature of the work, the locality and the cost of labour and materials prevailing at the time when the work is executed.

(2) Within the period of one month beginning with the day of service of the said account the adjoining owner may serve on the building owner a notice stating any objection he may have thereto and thereupon a dispute shall be deemed to have arisen between the parties.

(3) If within that period of one month the adjoining owner does not serve notice under subsection (2) he shall be deemed to have no objection to the account.

Settlement of account.

14.- (1) All expenses to be defrayed by an adjoining owner in accordance with an account served under section 13 shall be paid by the adjoining owner.

(2) Until an adjoining owner pays to the building owner such expenses as aforesaid the property in any works executed under this

Act to which the expenses relate shall be vested solely in the building owner.

Miscellaneous.

15.- (1) A notice or other document required or authorised to be served under this Act may be served on a person;

(a) by delivering it to him in person;

(b) by sending it by post to him at his usual or last-known residence or place of business in the United Kingdom; or

(c) in the case of a body corporate, by delivering it to the secretary or clerk of the body corporate at its registered or principal office or sending it by post to the secretary or clerk of that body corporate at that office.

(2) In the case of a notice or other document required or authorised to be served under this Act on a person as owner of premises, it may alternatively be served by –

(a) addressing it "the owner" of the premises (naming them), and

(b) delivering it to a person on the premises or, if no person to whom it can be delivered is found there, fixing it to a conspicuous part of the premises.

Offences

16.- (1) If –

(a) an occupier of land or premises refuses to permit a person to do anything which he is entitled to do with regard to the land or premises under section 8(1) or (5); and

(b) the occupier knows or has reasonable cause to believe that the person is so entitled,

the occupier is guilty of an offence.

(2) If –

(a) a person hinders or obstructs a person in attempting to do anything which he is entitled to do with regard to land or premises under section 8 (1) or m(5); and

(b) the first-mentioned person knows or has reasonable cause to believe that the other person is so entitled,

the first-mentioned person is guilty of an offence.

(3) A person guilty of an offence under subsection (1) or (2) is liable on summary conviction to a fine of an amount not exceeding level 3 on the standard scale.

Recovery of sums.
17. – Any sum payable in pursuance of this Act (otherwise than by way of fine) shall be recoverable summarily as a civil debt.

Exception in case of Temples etc.
18.- (1) This Act shall not apply to land which is situated in inner London and in which there is an interest belonging to –

(a) the Honourable Society of the Inner Temple,

(b) the Honourable Society of the Middle Temple,

(c) the Honourable Society of Lincoln's Inn, or

(e) the Honourable Society of Gray's Inn.

(2) The reference in subsection (1) to inner London is to Greater London other than the outer London boroughs.

The Crown.
19. – (10 This Act shall apply to land in which there is –

(a) an interest belonging to Her Majesty in right of the Crown,

(b) an interest belonging to a government department, or

(c) an interest held in trust for Her Majesty for the purposes of any such department.

(2) This Act shall apply to –

(a) land which is vested in, but not occupied by, Her Majesty in right of the Duchy of Lancaster;

(b) land which is vested in, but not occupied by, the possessor for the time being of the Duchy of Cornwall.

Interpretation.
20.- In this Act, unless the context otherwise requires, the following expressions have the meanings hereby respectively assigned to them

"adjoining owner" and "adjoining occupier" respectively mean any owner and any occupier of land, buildings storeys or rooms adjoining those of the building owner and for the purposes only of section 6 within the distances specified in that section;

"appointing officer" means the person appointed under this Act by the local authority to make such appointments as are required under section 10(8);

"building owner" means an owner of land who is desirous of exercising rights under this Act;

"foundation", in relation to a wall, means the solid ground or artificially formed support resting on solid ground on which the wall rests;

"owner" includes –

(a) a person in receipt of, or entitled to receive, the whole or part of the rents or profits of land;

(b) a person in possession of land, otherwise than as a mortgagee or as a tenant from year to year or for a lesser term or as a tenant at will;

(c) a purchaser of an interest in land under contract for purchase or under an agreement for a lease, otherwise than under an agreement for a tenancy from year to year or for a lesser term;

"party fence wall" means a wall (not being part of a building) which stands on lands of different owners and is used or constructed to be used for separating such adjoining land of one owner the artificially formed support of which projects into the land of another owner;

"party structure" means a party wall and also a floor partition or other structure separating buildings or parts of buildings approached solely by means of separate staircases or separate entrances;

"party wall" means –

(a) a wall which forms part of a building and stands on lands of different owners to a greater extent than the projection of any artificially formed support on which the wall rests: and

(b) so much of a wall not being a wall referred to in paragraph (a) above as separates buildings belonging to different owners;

"special foundations" means foundations in which an assemblage of beams and rods is employed for the purpose of distributing any load; and

"surveyor" means any person not being a party to the matter appointed or selected under section 10 to determine disputes in accordance with the procedures set out in this Act.

Other statutory provisions.

22. – (1) The Secretary of State may by order amend or repeal any provision of a private or local Act passed before or in the same session as this Act, if it appears to him necessary or expedient to do so in consequence of this Act.

(2) An order under subsection (1) may –

(a) contain such savings or transitional provisions as the Secretary of State thinks fit;

(b) make different provision for different purposes.

(3) The power to make an order under subsection (1) shall be exercisable by statutory instrument subject to annulment in pursuance of a resolution of either House of Parliament.

General

22. – (1) This Act may be cited as the Party Wall etc. Act 1996.

(2) This Act shall come into force in accordance with provision made by the Secretary of State by order made by statutory instrument.

(3) An order under subsection (2) may –

(a) contain such savings or transitional provisions as the Secretary of State thinks fit;

(b) make different provision for different purposes.

(4) This Act extends to England and Wales only.